W9-CMH-827

THE CHEMISTS AND THE WORD

THE CHEMISTS AND THE WORD

The Didactic Origins of Chemistry

OWEN HANNAWAY

The Johns Hopkins University Press · Baltimore and London

Manufactured in the United States of America

The Johns Hopkins University Press, Baltimore, Maryland 21218
The Johns Hopkins University Press Ltd., London

Library of Congress Catalog Card Number 74-24380
ISBN 0-8018-1666-1

Library of Congress Cataloging in Publication data
will be found on the last printed page of this book.

FOR CAROLINE

CONTENTS

PREFACE

This book is a contribution to the study of an important episode in the history of chemistry. It seeks to explain the emergence of chemistry as an integral and distinctive discipline around the turn of the seventeenth century. This development is most visibly manifest in the appearance of numerous specialized textbooks on chemistry which purport to teach this subject as a distinctive art. The number of these books increases throughout the century, and some few, like Jean Beguin's *Tyrocinium Chymicum* (first edition Paris, 1610) and Nicolas Lemery's *Cours de Chymie* (Paris, 1675), achieve extraordinary popularity, judging from the number of reprints and editions produced. The vast majority of these works grew out of actual courses taught by their authors both privately and institutionally, including a few which were products of university courses in chemistry. The chemical teacher, the chemical course, and the chemical textbook are amongst the most notable and pervasive manifestations of scientific didactic in the seventeenth century. The implications of this are apparent, even though historians have tended to gloss over their import: chemistry achieved in the seventeenth century sufficient coherency and identity to sustain a vigorous and ongoing didactic tradition, which gained for it remarkably early recognition as a unique discipline and activity. This was achieved in a subject which is generally regarded as only having come to scientific maturity in the late eighteenth century.

The origins of these formal expressions of disciplinary identity remain somewhat obscure. While the content of these textbooks has been fairly exhaustively mined for what they reveal about the cumulative development of chemical technique and preparation, the circumstances which prompted them are far from clear. In recent years the attention of historians of early chemistry has been diverted from the narrow confines of disciplinary history by the notable new research which has been conducted on the Paracelsian movement by such scholars as Walter Pagel, Allen Debus, P. M. Rattansi, Charles Webster, and others. This has demonstrated in a quite new way the pervasiveness and influence of the chemical ideology which was the core of the Paracelsian movement. It has come to be recognized that Paracelsianism, together with related Hermeticism, was a powerful and influential world view which touched many aspects of sixteenth- and seventeenth-century culture, not least its science. This chemical philosophy sought to comprehend natural phenomena, and by analogy, supernatural ones, in terms of chemical reactions and processes. It was an alternative to the rational Aristotelian and scholastic philosophy which it sought to overthrow and to the mechanical philosophy which came to be its chief competitor. In addition its advocates have been shown to be amongst the most vociferous critics of established social and academic institutions, prominent in the cry for university reform. In short, this movement is now seen as a much more significant ingredient in the intellectual and social ferment which engulfed the seventeenth century than had previously been suspected.

In the light of this new knowledge, it might have been expected that the parallel emergence of chemistry as an integral discipline would have been clarified. The irony, however, is that this deeper understanding of the implications of the chemical philosophy, while it has done much to expand our appreciation of the complexity of the scientific revolution as a whole, has only pointed up some perplexities regarding the origins of the discipline of chemistry. When the main contribution of Paracelsianism to chemistry was seen simply in terms of its advocacy of chemically prepared medicaments, the textbook tradition could be seen as the formalization of instruction in the techniques and prescriptions of the new therapeutics. What remains puzzling, however, is the obvious divorce and independence of the textbooks from Paracelsian ideology. Did chemistry in its didactic form simply abstract the more positivistic elements of Paracelsian prescription while abandoning the mystical and alchemical theory which underlay it? Or is there some other source of inspiration for its development?

Investigation into the origins of chemistry as a distinctive subject of instruction appears to lead back to one text, the *Alchemia* of Andreas Libavius, published in 1597. This work has long been recognized as a landmark in chemical literature—the first comprehensive synthesis which seeks to integrate the techniques and preparations of all the chemical arts and crafts and to present chemistry as a subject worthy of study for its own sake. This recognition, however, has not been matched by an adequate comprehension of its motivation and intention. The author himself appears as an incongruous figure, an unlikely author of a major chemical textbook. He was for a large part of his life a *gymnasium* teacher whose interests extended beyond chemistry to include logic, theology, poetry, pedagogy, and philosophy. One fact about him is incontrovertible: he was avowedly anti-Paracelsian, in spite of the fact that he appeared to exploit Paracelsian recipes in his synthesis. Clearly an understanding of what Libavius's goals were in writing his textbook and, in particular, what his position was vis-à-vis the Paracelsians would illuminate greatly the transition from chemical philosophy to chemical didactic.

The key to an understanding of these problems was found in one of the many polemical writings of Libavius, that directed at the Paracelsian philosopher Oswald Croll. Croll is well known as the author of an influential text entitled the *Basilica Chymica* (1609), which set forth a concise and forceful statement of Paracelsian-Hermetic philosophy and included a comprehensive description of Paracelsian remedies. It is this latter part of the work which has received the bulk of attention from historians seeking a clear understanding of the particular contributions of Paracelsian chemical therapy to pharmacy. Libavius's critique, however, is directed at the theoretical section of Croll's work and reveals with particular clarity his objections and hostility to the whole Paracelsian ideology. There are also significant clues as to the ideas and inspiration for his own work, the *Alchemia.* Reading this critique, many of the apparently incongruous elements of Libavius's career and writings began to fall into place. What was revealed in this polemic with Croll was a fundamental clash of ideologies which ran much deeper than the status and provenance of chemistry. This clash was found to be rooted in differing and irreconcilable interpretations of the signification of the word and the role of language in the explication of nature. Hence the title of this book.

In an effort therefore to clarify not only Libavius's accomplishments but also the relationship of the emergence of chemistry as a discipline to the chemical philosophy of the Paracelsians, I have chosen to juxtapose a discussion of Oswald Croll and Andreas Libavius and to analyze their two principal works in the light of their underlying philosophies. This I believe will make clear that chemistry conceived as a discipline was something radically different from the chemical philosophy of the Paracelsians. In making this contrast I have taken some liberties with chronology. Libavius's textbook, the *Alchemia,* was published some twelve years before the *Basilica Chymica* of Oswald Croll; his critique of Croll was not published until near the end of his life, in 1615. However Croll's statement of the Paracelsian ideology and the response it elicited from Libavius point up so well the fundamental issues at stake that I have used Croll's *Basilica Chymica* as a foil against which to present my interpretation of Libavius's motives and accomplishments. This book thus emerges as a study of contrasts and conflict, not of evolution and continuity, which it is hoped will throw some new light on the origin of the discipline of chemistry.

In the development of the themes of this book I am particularly indebted to certain authors. In my analysis of Croll I was greatly aided by the insights provided by Michel Foucault in chapter 2 of his *The Order of Things,* entitled "The Prose of the World." It should be stated, however, that I have not consciously attempted to follow Foucault methodologically in an effort to uncover an archaeology of chemical knowledge; nor is the discontinuity I have depicted between Croll and Libavius to be identified with that which Foucault has claimed to detect between the sixteenth century and what he calls the Age of Classicism. Walter Ong's *Ramus, Method, and the Decay of Dialogue* was critical in crystallizing my perception of Libavius. My debt to this work is obvious throughout chapters 5 and 6, and I doubt that I could have made my way through so much of Libavius's writings without Ong's guidance. Neal W. Gilbert's *Renaissance Concepts of Method* also helped put Libavius's pedagogical ideas in their proper context. R. Hooykaas's brief study entitled *Humanisme, Science et Réforme: Pierre de la Ramée 1515-72* helped me understand the significance of Ramist ideas in relation to science.

No work which encompasses Paracelsus and the Paracelsians could fail to be influenced and aided by the writings of Walter Pagel. My footnotes most fully reveal my debt to him. For the theological dimensions of Paracelsian thought I found the writings of Kurt Goldammer a most valuable introduction, particularly his *Paracelsus: Natur und Offenbarung.* Likewise my explication of the Hermetic element in Croll's Paracelsianism would not have been possible without the pioneering studies of D. P. Walker and Frances Yates. To these and other authors cited who enabled me to appreciate the wider dimensions of this study in the history of chemistry I am deeply grateful.

Finally I should like to thank here those many individuals who have contributed in a more personal way to the evolution of this book. I owe a very special debt of gratitude to Dr. Andrew Kent, my former mentor in the chemistry department at the University of Glasgow, who first set me on the path of this study by placing in my hands a copy of Libavius's *Alchemia* and who guided my early research into the history of chemistry. I should also like to thank Professors J. M. Robertson, D. W. J. Cruikshank, and W. A. Smeaton for encouragement in those early Glasgow days. Professor Aaron J. Ihde kindly afforded me the opportunity to further my research in the history of science department at the University of Wisconsin. Harry Woolf introduced me to the stimulating environment of The Johns Hopkins University, where I have benefited in innumerable ways from the knowledge and experience of colleagues in many departments. To my friends and colleagues in the history of science department, William Coleman and Robert Kargon, I am especially grateful: they both provided patient support, understanding, and encouragement throughout my research and writing. I also had the benefit of the helpful advice and criticism of Camille Limoges, presently of the Université de Montréal, at a critical stage in the writing of this book. Dr. Owsei Temkin and Professor F. L. Holmes made several valuable suggestions which aided my revisions. I am responsible, however, for all interpretations and any errors which remain. I would also like to thank Anne Carey, who typed various drafts of the manuscript.

*The principal themes of Croll's exposition of chemical philosophy are illustrated in the plate opposite. The realm of the light of grace is depicted above the title, that of the light of nature below. The focal point of the former is the ineffable name of God (*Yahweh*) at the center of the equilateral triangle representing the Trinity of Father, Son, and Holy Spirit. The corresponding triangle in the light of nature has as its center the "adamic earth," from which Adam's body was created and which contained all the virtues of nature, thereby making man the center of the natural world. The apices of this latter triangle are denoted by the alchemical symbols of the three principles of mercury, sulphur, and salt, corresponding to the Persons of the Trinity. Other triads depicted in nature are those of the three kingdoms—animal, vegetable, and mineral; soul, body, and spirit; and the elements fire, water, and air. The three sciences of man are denoted as theological Cabala, medical alchemy, and astronomical magic. The zodiacal band of stars in nature corresponds to the orders of angels in the realm of grace. The kneeling figure above the furnace seeks light from the Hebrew letters of the divine name of Jesus, the Word Incarnate, the unique link between the light of nature and the light of grace. The lute opposite this figure is an allusion to the Orphic mysteries. The whole page is framed by portraits of notable alchemical philosophers with quotations from their writings.*

HERMES TRISMEGISTOS ÆGYPTI

GEBER ARABS

Cherubim

Seraphim

Throni

Dominationes

Potestates

Principatus

Virtutes

Archangeli

Angeli

LVX IN ACCESSA

In Sole et Sale Naturæ Sunt omnia.

D·O·M·A·

MORIENES ROMANVS

ROG. BACCHON ANGLUS

OSUALDI CROLLII VETERANI HASSI

BASILICA CHYMICA

CONTINENS,

Philosophicam propriâ laborum experientiâ

confirmatam descriptionem et usum Reme-

diorum Chymicorum Selectissimorum é

Lumine GRATIÆ et NATURÆ

DESUMPTORUM.

In fine libri additus est Autoris ejusdem Tractatus

Nouus DE SIGNATURIS Rerum Internis.

Occultum fiat mani: festum, et viceuersa

Per elementorum conversionem Ternarius purificatus fiat MONAS

R. LULLIUS HISPANUS

TH. PARACELSVS GERMANUS.

LUMINIS NATURÆ BONUM FINITUM

ORA ET LABORA

REPUTANS OMNIA VANA PRÆTER AMARE DEUM ET ILLI SOLI SERVIRE.

CABALA THEOLOGICA MAGIA ASTRONOMICA ALCHYMIA MEDICA

Ignis

ANIMA

Animalia

Terra

Vegetabilia

Mineralia

SPIRITUS

CORPUS

Aqua

Aer

Cum Igne tandem In gratiam redit Aqua.

Separate et ad maturitatem perducite.

Plate 1. Title Page of Oswald Croll, *Basilica Chymica* (1609). (Courtesy of the National Library of Medicine, Bethesda, Maryland.)

CHAPTER I

THE LIGHTS OF THE WORD

Amongst the German disciples of the Swiss medical reformer Paracelsus, Oswald Croll (ca. 1560–1609) stands out as the one who developed and systematized the chemical therapy of the master.[1] Croll's only published work, the *Basilica Chymica* of 1609,[2] was the summation of a lifetime's experience gained in a widely traveled career as physician, tutor, and minor diplomat. It is probable that he never saw the printed book, as he died early in the year of its publication.

Croll was born ca. 1560 in the town of Wetter in Hesse-Kassel. He was the third son of a mayor of this Lutheran town and was educated at the local reformed abbey school before proceeding in 1576 to the nearby University of Marburg. He later continued his studies at Heidelberg, Strasbourg, and Geneva and is thought to have graduated M.D. in 1582, but from where precisely is not known. His first position was that of tutor to the d'Esnes family in Lyons from 1583 to 1590. As a member of this household he traveled widely in France and Italy, mastering the languages of both countries. Then followed a similar position in Tübingen as teacher in the household of Count Maximilian von Pappenheim, which he abandoned in 1593 to travel for four years through the eastern part of the Holy Roman Empire. During this time

[1] For Croll's life, see Gerald Schröder, "Oswald Croll," *Pharm. Ind.* 21 (1959): 405–8, and the same author's much briefer article in *Dictionary of Scientific Biography,* "Crollius, Oswald" (New York: Charles Scribner's Sons, 1971–), vol. 3, pp. 471–72. Croll's contributions to chemiatric pharmacy are discussed in Robert Multhauf, "Medical Chemistry and 'The Paracelsians,' " *Bull. Hist. Med.* 28 (1954): 101–25; and Gerald Schröder, "Studien zur Geschichte der Chemiatrie," *Pharm. Zeit.* 111, no. 35 (1966): 1246–51.

[2] *Basilica Chymica, continens philosophicam propria laborum experientia confirmatam descriptionem & usum remediorum chymicorum selectissimorum è lumine gratiae et naturae desumptorum. In fine libri additus est eiusdem Autoris Tractatus novus de Signaturis Rerum Internis* (Frankfort, 1609), hereinafter to be referred to as *Basilica Chymica.*

he earned his living as a physician. He had a medical practice in Prague from 1597 to 1599, after which he moved into partnership with Dr. Johann Berger at Brno in Moravia. In 1602 both physicians returned to Prague, where Croll lived until his death.[3]

As an inhabitant of the imperial city Croll moved on the fringes of the distinguished circle of "occult" physicians and philosophers which surrounded the court of Emperor Rudolf II, who occasionally consulted Croll.[4] However, for most of the time he was in Prague, Croll was in the pay of the emperor's arch political enemy, Prince Christian I of Anhalt-Bernburg. This accomplished Calvinist prince, who from 1595 governed the Upper Palatinate, emerged in the 1590s as the political and military leader of the "corresponding" Protestant princes against the House of Hapsburg.[5] Croll apparently met Christian in 1598, at which time he cured him of a seemingly mortal disease. As an expression of his gratitude the prince appointed Croll his personal physician, although the latter continued to reside in Prague. This had certain political advantages for Anhalt, who used Croll for delicate diplomatic negotiations in and around the imperial city in furtherance of his project for an Evangelical Union of Protestant Princes.[6] In return for these services Christian provided financial support for Croll's chemical researches. The Preface of the *Basilica Chymica* is dedicated to this renowned champion of Protestant Europe.[7]

The *Basilica Chymica* contains a most comprehensive treatment of the chemical techniques and preparations of the new spagyric phar-

[3]The preceding biographical information is taken from Schröder's "Oswald Croll," in *Pharm. Ind.* and from his article in *D.S.B.* Croll himself makes passing reference to episodes in his life throughout the *Basilica Chymica.*

[4]See Schröder, *D.S.B.*, vol. 3, p. 471. Croll himself mentions treating Rudolf in the preface to the *Tractatus novus de Signaturis* (p. 4). The intellectual and artistic activities of the court of the emperor at Prague are brilliantly discussed in R. J. W. Evans, *Rudolf II and His World: A Study in Intellectual History, 1576–1612* (Oxford: Clarendon Press, 1973). Croll is specifically mentioned in several places in this work.

[5]Anhalt's policy was directed at breaking the hold of the House of Austria on the imperial throne. Key elements in that policy were winning the support of the estates of Bohemia and developing a military alliance with Henry IV of France in support of the United Provinces. French influence, political and cultural, was dominant at Anhalt's court. See A. W. Ward, "The Empire under Rudolph II," in *The Cambridge Modern History*, ed. A. W. Ward, G. W. Prothero, and Stanley Leathes, 14 vols. (New York and London: Macmillan, 1902–12), 3: 696–735, esp. pp. 712–35; also Frances A. Yates, *The Rosicrucian Enlightenment* (London and Boston: Routledge & Kegan Paul, 1972), pp. 27–28 and 36–37.

[6]For Croll's rôle in these negotiations see Schröder, "Oswald Croll," p. 407.

[7]Croll specifically acknowledges his debt to Anhalt at the end of the *Admonitory Preface* to the *Basilica Chymica.*

macy. It represents much more than an elaboration or commentary on the preparations to be found in the corpus of Paracelsus himself; it is a thoroughgoing exploitation of the potential in chemical technique and materials for the extension of the Paracelsian pharmacopoeic roster. Furthermore, its descriptions of preparations are clear and concise, unlike those of Paracelsus himself, and can be readily interpreted in modern terms. For these reasons, the preparative section of Croll's *Basilica Chymica* has been the focus of recent study of the evolution of chemiatric therapy and its impact on the history of Western pharmacy. Robert Multhauf has described Croll's work as the best example of "the fully developed chemical pharmacopoeia—that is, the pharmacopoeia of medical chemistry on the eve of its general acceptance as a part of legitimate medicine."[8] Croll's main contribution to medical chemistry, according to Multhauf, was the systematic use of the mineral acids in the preparation of saline remedies derived from such traditional alchemical substances as tartar, corals, pearls, antimony, mercury, and gold. Gerald Schröder, in his extensive studies on the influence of chemistry on the German pharmacopoeias, has confirmed and strengthened this estimate.[9] Several of these chemiatric preparations originating with Croll found a permanent place in standard pharmacopoeias until well into the nineteenth century.[10]

Whereas Schröder and Multhauf have concentrated on the preparative section of the *Basilica Chymica* in an effort to assess Croll's contribution to the Paracelsian reform of pharmacy, this section is only one third of the work. It is preceded by a long *Admonitory Preface,* of over one hundred pages in the original edition, which provides a synthesis of Paracelsian-Hermetic theories of medicine and an interpretation of the rôle of the physician by Croll; and it is followed by a separate treatise on the doctrine of signatures. Multhauf, who claims to see a growing divorce between Paracelsian theory and practice in the work of the early chemiatrists, fails to detect a relationship between the

[8]Multhauf, "Medical Chemistry and 'The Paracelsians,' " p. 110.

[9]In addition to Schröder's "Studien zur Geschichte der Chemiatrie," see also Schröder, *Die pharmazeutische-chemischen Produkte deutscher Apotheken im Zeitalter der Chemiatrie,* Publication of the History of Pharmacy Seminar of the Braunschweig Technische Hochschule, (Bremen, 1957), in which Schröder examines in some detail the appearance of chemiatric preparations in the German *Arzneitaxen* and the pharmacopoeias of the sixteenth and seventeenth centuries and gives a chemical analysis of their composition. For an overall assessment of Paracelsus's influence on pharmacy in the same period, see Wolfgang Schneider, "Die deutschen Pharmakopoën des 16. Jahrhunderts und Paracelsus," *Pharm. Zeit.* 106 (1961): 3–15; and "Der Wandel des Arzneischatzes im 17. Jahrhundert und Paracelsus," *Sudhoffs Archiv* 45 (1961): 201–15.

[10]Multhauf, "Medical Chemistry and 'The Paracelsians,' " p. 116.

theoretical introduction and the practical section which follows.[11] Certainly the *Admonitory Preface* does not provide any obvious blueprint for the ensuing practical recipes. However, it does contain an intense polemical justification for the new Paracelsian medicine and, *inter alia,* a plea for the involvement of the new breed of physician with chemical remedies and their preparation. All three sections of Croll's *Basilica Chymica* are important for an understanding of the mental outlook underlying the new medicine, which regarded chemistry as an integral part of the physician's training and thereby sought to incorporate it into the medical curriculum. If we are ever to understand the contribution of the Paracelsians to the development of chemistry and medicine, we must seek to understand them on their own terms. In retrospect we may recognize the importance and significance of their additions to the list of pharmaceuticals; but to regard them simply as exploiting systematically the physical and chemical properties of a restricted range of inorganic materials for therapeutic ends not only represents just one facet of their activity but also perpetuates a misleading view of their goals and objectives. Even a cursory reading of the *Admonitory Preface* reveals that Croll would have been disconcerted to learn that his reputation rested on his skillful manipulation of chemical techniques and material substances. The transformations of mere matter did not concern him; his goal was the control and manipulation of the spiritual, cosmic forces of nature which lay encapsulated within the useless dross of matter.

The principal sources of the ideas which stimulated and supported Croll's advocacy of chemical therapeutics were of course the writings of Paracelsus.[12] In essence Croll's message is the same as that of Paracelsus. He offers similar criticisms of classical medical theory and practice; he calls for a renewal of medicine based upon the individual "experience" of the physician; and he sets forth a similar program for a heightened sense of social and religious responsibility on the part of the medical practitioner. Croll never wavers in his loyalty to Paracelsus, and the Swiss reformer holds undisputed pride of place in his pantheon of great doctors and profound illuminati. But just as Croll clarified and

[11]Ibid., p. 108.

[12]The secondary literature on Paracelsus is as extensive as that on any figure in the history of medicine. Three major works served as the basis for the present study: Walter Pagel, *Paracelsus: An Introduction to Philosophical Medicine in the Era of the Renaissance* (Basel and New York: S. Karger, 1958); idem, *Das Medizinische Weltbild des Paracelsus: Seine Zusammenhänge mit Neuplatonismus und Gnosis* (Wiesbaden: Franz Steiner Verlag GmbH, 1962); and Kurt Goldammer, *Paracelsus: Natur und Offenbarung* (Hanover: Theodor Oppermann Verlag, 1953). Pagel, *Paracelsus,* pp. 31–35, provides a brief but useful guide through the literature.

reordered the prescriptions of Paracelsian therapy, so also did he work out his own unique synthesis of Paracelsian doctrine, which reflected the religious, political, and cultural circumstances in which he lived. Paracelsus was, after all, a wandering outcast of the early sixteenth century, a contemporary of Erasmus and Luther who, although strongly influenced by the early stirrings of the Renaissance and Reformation in northern Europe, still had deep roots, both spiritually and intellectually, in the Middle Ages.[13] Croll on the other hand occupied a secure and privileged position in one of the major cultural and political centers of a Europe which had experienced the full flowering of the Renaissance but which in the aftermath of the Reformation was increasingly fragmenting politically along rigid confessional lines and was hellbent down the path leading to the catastrophe of the Thirty Years' War.

The sixty-eight years between the death of Paracelsus in impoverished obscurity at Salzburg in 1541 and the appearance of the *Basilica Chymica,* complete with imperial privilege, in 1609 certainly left their mark on Croll's work. The transformations which the tenets of Paracelsianism underwent in Croll's hands are significant but subtle; their precise character will only become apparent when we have analyzed the *Admonitory Preface* in some detail. In the remainder of this chapter, however, we shall examine some of the broad themes which are woven into Croll's restatement of Paracelsian medicine. Most of these are inherent in Paracelsus's own writings, but Croll brings them into sharper focus in the light of intervening cultural and religious developments. While it is impossible to be definitive about the changes, principally because Paracelsus was never one to admit to "intellectual" influences and still less one to document his "sources," certain factors stand out: namely, Croll's location of Paracelsus in the mainstream of protestant mystical theology of the sixteenth century; his integration of Paracelsian medicine into the fashionable Hermetic doxography of the early seventeenth century; and finally the seal of Cabalistic exegesis which he sets on his whole synthesis. While none of this does real violence to any of Paracelsus's teachings, except, perhaps, in the sphere of theological interpretation, it does provide a more coherent articulation of Paracelsian theory in the context of early seventeenth-century culture.

Fundamental to an understanding of the Paracelsian theory of knowledge is the doctrine of the two lights—the light of nature and the light of grace. The symbol of light as the source and vehicle of intellectual knowledge and spiritual awareness has deep and ancient roots in Western consciousness. In the classical tradition it was a

[13] Goldammer, *Paracelsus,* p. 59.

fundamental concept of the Pythagoreans and the Neoplatonists. Thence the image was absorbed into Christian symbolism, most notably in the writings of St. John and St. Paul, whence it became one of the commonplaces of medieval scholastic theology. The light-symbol has been exploited particularly in philosophies which sought to comprehend rational knowledge in the context of religious experience—sometimes to relate the two, other times to contrast them. One of the minor ironies of Western intellectual history is that the rational philosophers of the eighteenth century made this particular symbol peculiarly their own.

The Renaissance brought a remarkable efflorescence of light-symbolism in philosophical and religious writing. For Marsilio Ficino and Pico della Mirandola it was a central image in their restatement of a Platonic and Neoplatonic Christianity; Nicholas of Cusa exploits the lightness-darkness antinomy in his philosophical and mystical paradoxes; and light plays an important rôle in Agrippa von Nettesheim's theory of magic. But as Kurt Goldammer has argued in a recent article on the pervasiveness of the *lichtsymbolik* from the fifteenth through the seventeenth centuries, the most original, comprehensive, and influential use of the metaphor was made by Paracelsus.[14]

Following Goldammer's careful analysis, we see that Paracelsus's concept of the two lights evolved gradually throughout the course of his lifetime. This evolution represented a resolution of a basic tension in Paracelsus's thought and in his perception of his mission. In search of his true vocation, he felt himself drawn in opposite directions for a large part of his life: one way led along a path in the search for God in His creation and a fulfillment of his Christian obligation by curing men's bodily afflictions through the means God had provided in nature; the other way lay along a more exalted path, which sought God in His transcendental relationship with man through mystical experience and a calling to heal men's souls through the preaching of the Word.[15] In this regard it should be recalled that Paracelsus left almost as extensive a *nachlass* of theological writings as he did of medical works. The one path was illuminated by the light of nature, the other by the light of grace; and each path had its corresponding guide—the Book of Nature and the Book of Scripture.

In the earlier writings of Paracelsus, dating from around 1520,

[14]Kurt Goldammer, "Lichtsymbolik in philosophischer Weltanschauung, Mystik und Theosophie vom 15. bis zum 17. Jahrhundert," *Studium Generale 13* (1960): 670–82; this article includes a discussion of Nicholas of Cusa, Agrippa, Paracelsus, Franck, Weigel, and Boehme.

[15]These aspects of Paracelsus's career are best brought out in Goldammer, *Paracelsus*, esp. pp. 19–26, 53–59, and 68–74.

this tension is not yet manifest. Paracelsus is confident in his vocation as a physician and in his ability to reform medicine on the basis of the light of nature, which he equates with the light of the intellect belonging to every man from birth. This light is present within man, but it is part of the "eternal" light which descends into his soul from the angels via the stars. The "eternal" light is the source of the natural light. This formulation, as Goldammer points out, is very similar to the Neoplatonic syncretism of Agrippa von Nettesheim.[16]

By the end of the 1520s, however, Paracelsus seems to have undergone an intellectual and spiritual crisis. Both his medical and theological writings demonstrate that he no longer has unlimited confidence in the light of nature. He still values this light above the authority of Aristotle, but he contrasts it with the "better" light of God, which is the Holy Ghost. In his commentary on the Psalms, dating from this period, he states that man's arts and crafts come not from the light of nature but from God.[17] The same tension is manifest in his medical writings from this period. Here the light of nature is equated no longer with man's intellectual capacities but with the active powers of the physician's art, derived from "experience." Here again, the light of nature, which teaches physicians, is sharply distinguished from the Holy Ghost, who teaches the faithful. Paracelsus does however refer to the Holy Ghost as the spark (*der Anzunder*) of the light of nature.[18] These uncertain and somewhat vague distinctions betoken a crisis of self-doubt in the efficacy and value of his vocation as a physician, which is only resolved in his great philosophical and religious synthesis, the *Philosophia Sagax* of 1537.[19] In this work Paracelsus recognizes and reconciles himself to his own limitations as regards his healing mission, by abandoning the antinomy of the eternal light and the natural light and making them complementary. Both lights are divine. The light of nature is the peculiar gift of God the Father which testified to God's beneficence and mercy in His creation. It is the light which guided all men, including the heathens, between the Fall and the Incarnation of Christ. It enabled man to employ the divine powers which God had reposited in His creation for the alleviation of his temporal needs and

[16]Goldammer, "Lichtsymbolik," p. 676.

[17]Ibid., p. 677.

[18]Ibid., p. 676.

[19]*Astronomia Magna oder die ganze Philosophia Sagax der grossen und kleinen Welt*, in *Theophrast von Hohenheim, genannt Paracelsus, Sämtliche Werke. I. Abteilung: Medizinische, naturwissenschaftliche und philosophische Schriften*, ed. Karl Sudhoff, 14 vols. (Munich: R. Oldenbourg, 1922–33), 12:1–444, hereinafter to be referred to as *Paracelsus, Sämtliche Werke*. On the significance of this work see Goldammer, "Lichtsymbolik," p. 677.

afflictions. When Christ was born the eternal light entered the world and greatly outshone the natural light. But it did not extinguish it; rather, the eternal light—the light of grace—complemented and fulfilled the light of nature. The Christian was enjoined to make use of both lights, employing the powers of nature according to the spirit and the ends revealed by the light of grace. Although the prophets, apostles, and Christ manifested great powers through the eternal light, their accomplishments did not negate the more humble efforts of those who continued to operate through the light of nature. It was in this spirit that Paracelsus reconciled himself to his own mission:

> . . . and though I myself often write in the manner of the heathen, I am a Christian and must recognize that the inferior wisdom must yield to the greater. The wisdom of Christ is more profound than that of nature, consequently a prophet or an apostle must be held in higher esteem than an astronomer or a physician; and it is better to prophesy from God than through astronomy, it is better to cure through God than by means of herbs. The message of the prophets is without an error. The apostles have power to cure the sick and raise the dead, and likewise their works are without flaw. Who then can doubt that astronomy and its light have paled beneath the light of Christ? Nevertheless it is our duty to say that the sick need a physician, while few need an apostle; similarly many forecasts must be made by the astronomer and not by the prophets. Thus each has his part—the prophet, the astronomer, the apostle, and the physician. . . . For the Father has set us in the light of nature, and the Son in the eternal light. Therefore it is indispensable that we should know them both.[20]

Croll's exposition of Paracelsus is developed within the theme of the reconciliation of the two lights as set out in these late writings. When we turn to Croll we see again that the two lights are complementary. In this passage, and in all subsequent quotations, I have used the English translation of the *Admonitory Preface* made by Henry Pinnell in 1657, which captures magnificently the religious fervor of the work in a vigorous prose.[21]

[20]This quotation is from one of the drafts made by Paracelsus for the *Philosophia Sagax*, printed in *Paracelsus, Sämtliche Werke*, 12:496–97. The English translation, by Norbert Guterman, in Jolande Jacobi, ed., *Paracelsus: Selected Writings*, Bollingen Series 28, 2d ed., (New York: Pantheon Books, 1958), pp. 154–55, is from this text.

[21]*Philosophy Reformed and Improved in Four Profound Tractates. The I. Discovering the Great and Deep Mysteries of Nature: By that Learned Chymist and Physitian Osw: Crollius. The Other III. Discovering the Wonderfull Mysteries of the Creation, By Paracelsus: Being His Philosophy to the Athenians. Both made English by H. Pinnell, for the increase of Learning and true Knowledge* (London, 1657). This contains a complete translation of the *Admonitory Preface* of Croll's *Basilica Chymica* only. Unless stated otherwise, all subsequent references to the *Admonitory Preface* will be to this translation, and the original spelling has been retained. Henry Pinnell was an important antinomian author of the civil war period in

> The two Lights are well known, within which are all thing[s], without which is nothing, and no perfect knowledge of any thing. The Light of Grace, begetteth a true Theologer, yet not without Phylosophy: The Light of Nature, which is the Treasury of God confirmed in the Scriptures, maketh a true Phylosopher, yet not without Theologie, which is the Foundation of true Wisdome. The works of God are bipactite (*sic*): Philosophy comprehendeth the works or way of Nature; Theologie onely knoweth the works and way of Christ: In these two wayes we are to walk and spend our short time, that we may die in Peace and Joy. Hence it is plaine that every true Theologer is a Phylosopher, and every true Phylosopher is a Theologer.[22]

Thus for Croll Paracelsus is not simply a medical reformer but also a religious reformer. From his early seventeenth-century perspective Croll sees Paracelsus as one of the founders of the tradition of sixteenth-century mystical enthusiasm which stretched from Hans Denk (d. 1527), Caspar Schwenckfeld (1490–1561), and Sebastian Franck (1499–1542) to Valentin Weigel (1533–88) and found its culmination, in the seventeenth century at least, in the writings of Jacob Boehme (1575–1624).[23] The members of this tradition, although differing widely in their individual tenets, shared a common belief in the essentially inner spiritual nature of Christian experience and a common opposition to the emergent doctrinal orthodoxies of Protestantism. Their concept of the Church was one of an invisible body of true believers, as opposed to a visible, organized, and state-supported body of adherents, distinguished by their acceptance of a prescribed formulary of beliefs, ritual, and discipline. The inclusion of Paracelsus in this company is by no

England and a chaplain in the New Model Army until his resignation sometime prior to 1648. Like many antinomians in the Interregnum, Pinnell sought a more quietist spirituality and found refuge in the doctrines of Paracelsus and Croll. See "The Translators Apology, Caution and Retraction" to *Philosophy Reformed;* and Gertrude Huehns, *Antinomianism in English History: With special reference to the period 1640–1660* (London: The Cresset Press, 1951).

22 *Admonitory Preface,* p. 135.

23 For the mystical tradition in reformed theology up to Valentin Weigel, see Steven E. Ozment, *Mysticism and Dissent: Religious Ideology and Social Protest in the Sixteenth Century* (New Haven and London: Yale University Press, 1973), which contains an extensive and up-to-date bibliography. Paracelsus is discussed in the context of this movement in Alexandre Koyré, *Mystiques, spirituels, alchimistes: Schwenckfeld, Seb. Franck, Weigel, Paracelse* (Paris: A. Colin, 1955). Koyré's assessment of Weigel is subject to revision in the light of subsequent scholarship (see Ozment, *Mysticism and Dissent,* p. 210, n. 29), but the whole work remains of great value. Other useful introductions to the subject in English are Rufus M. Jones, *Spiritual Reformers in the 16th and 17th Centuries* (London: Macmillian, 1928); and George H. Williams, *The Radical Reformation* (Philadelphia: The Westminster Press, 1962). The best overall account of Boehme remains Alexandre Koyré, *La Philosophie de Jacob Boehme* (Paris: J. Vrin, 1929). See also Hans Aarsleff's article in *D.S.B.*, "Boehme, Jacob," vol. 2, pp. 222–24, for an extensive bibliography.

means wholly inappropriate. Paracelsus had much in common with the early *Schwarmgeister* and particularly with Sebastian Franck, whom he probably met on more than one occasion.[24] It is also significant that such later figures in the movement as Weigel and Boehme incorporated large elements of Paracelsus's own philosophy into their *Weltanschauung.* But it should not be forgotten that Paracelsus died a Catholic and that his spirituality was deeply rooted in the Catholic tradition and its liturgy. This is most clearly shown in the importance he assigned to the mother of God in his theological writings.[25] It is probable that Paracelsus would not have recognized some of the doctrinal presuppositions which later mystics with affinities to the Protestant tradition brought to the interpretation of his writing. As we shall see in a later chapter, Croll himself reinterprets some of Paracelsus's religious views in a context quite alien to the latter's spirituality.

As to Croll's own confessional allegiance, it seems most probable that he was, outwardly at least, a Calvinist. The travels of his early *Wanderjahren* from Marburg to Heidelburg, Strasbourg, and Geneva might well indicate a parallel theological journey from Lutheranism to Calvinism; his sojourn in Lyons and most significantly his service to Christian I of Anhalt-Bernburg point to a similar conclusion. We shall adduce further evidence in support of Calvinist leanings in the next chapter. Nevertheless, whatever his denominational persuasion might have been, Croll is quite explicit in the *Admonitory Preface* in regard to his admiration of the proponents of the "invisible theological philosophy," which he describes as that ". . . *Intellectuall* school of Pentecost, in which the Prophets, and Apostles, and all truly learned men walking in the Life and steps of Christ, have been taught and learned without labour and toyle. . . ."[26] Croll considered Paracelsus the founder of this school and included amongst later disciples Valentin Weigel,[27] which would seem to indicate that Croll had access to some of Weigel's

[24]See Pagel, *Paracelsus,* pp. 40–44; and Goldammer, *Paracelsus,* pp. 94–99 and 102–11.

[25]Paracelsus was by no means an orthodox Catholic, as Pagel and Goldammer have pointed out. His mystical and social views made him a severe critic of all forms of institutional Christianity. Nevertheless his spirituality remained deeply rooted in the Catholic tradition. See the comments of Jacobi in his introduction to *Paracelsus: Selected Writings,* p. lxvii.

[26]*Admonitory Preface,* p. 136.

[27]"After Paracelsus others attempted this study [i.e., theological philosophy] following the same strait and compendious tract, most holy godly men of blessed and honourable memory and most sound both in innocence and learning, such as Paulus Brawn of Norimberge, Velentinus Weigelius, and Petrus Winzius . . . " (*Admonitory Preface,* p. 135). I have not been able to identify Paul Brown or Peter Winzius.

manuscripts, as the heterodox writings of this outwardly orthodox Lutheran pastor who died in 1588 did not begin to appear in print until the year of Croll's death.[28]

The whole of Croll's *Basilica Chymica* must be understood in the context of this religious background. To neglect this basic element of the Paracelsian mentality while concentrating on the "proto-scientific" content of its teachings and practice is to distort the whole significance of the movement and to ignore its fundamental motivation. The Paracelsian reform of philosophy and medicine was preeminently a religious and ethical reform which drew deeply from spiritual experience. As Croll puts it:

> The true Physick whereof by the Devine assistance I intend here to treat, is the mear gift of the most high God; it is not to be sought for or learned from the Heathens, but from God alone, the Ancient of days, the Father of Lights, who cannot erre, the One onely Governour of the supream Universe.[29]

Although the doctrine of the two lights had a venerable history in Christian thought, it took on a new dimension during the Renaissance, in the form of an association of magical powers with the two lights. Illumination led not only to a purely passive knowledge of God's spiritual designs for mankind in Scriptures and a pious appreciation of His wisdom through the study of nature; but through this knowledge man could acquire spiritual and natural powers. The theory of the two lights became a springboard to action. Man illuminated by the light of nature could gain knowledge and control of the natural forces with which God had endowed His universe; man illuminated by the light of grace could attain to mystical experience and exercise supernatural powers.

Towards the end of the fifteenth century magic surfaced from the dregs of medieval necromancy with a new and dazzling patina of scholarly respectability compounded of a formidable mixture of philosophical sophistication, aesthetic sensibility, exegetical profundity, and classical erudition. The primary locus of this development was the Platonic Academy of Florence, with its two principal magi, Marsilio Ficino (1433–99) and Giovanni Pico della Mirandola (1463–94). This rehabilitation of magic was an integral part of these Florentines' search for a new, and hence ancient, Christian gnosis which would incorporate Plato and Neoplatonism in its doxography. Pico was the more ambi-

[28] Winfried Zeller, *Die Schriften Valentin Weigels: Eine literarkritische Untersuchung* (Berlin, 1940), p. 68.

[29] *Admonitory Preface*, p. 22.

tious in his syncretism, seeking to include Aristotelianism and scholasticism, as well as Persian, Arabic, Hindu, and Jewish gnostic sources, in his synthesis, and was more daring in his mystical and magical reach. Central elements in this program were the rediscovery of the ancient Egyptian wisdom and magic incorporated in the Revelations of the supposedly preclassical sage Hermes Trismegistus and the revival of the equally ancient Jewish mystical gnosticism and its magic, the Cabala. The daring flights of fancy of Marsilio and Pico are now fairly well-known, thanks in large part to the scholarly efforts of D. P. Walker and Frances Yates.[30]

One of the principal preoccupations of the Florentine magical school was to demonstrate that their magic was a "good" magic, one which did not dabble with the supernatural powers of evil. In this respect the older Ficino was much more cautious. The vehicle of his magic was the *spiritus mundi,* a pneuma of finest corporeity which circulated through the cosmos acting as a vivifying principle in the manner of a Neoplatonic emanation. This *spiritus* was intimately connected in Ficino's mind with the vital spirits, which in classical orthodox physiology were conveyed through the organs of the body and gradually refined to produce the vital, sensory, and rational functions of the human species. Ficino's magic was set in a largely medical context—he regarded himself primarily as a priest-physician—and was directed at drawing into the body beneficient forms of *spiritus* which could supplement a deficiency due either to the individual's temperament or to his habit of life. To Ficino, the planets were important sources of therapeutic spirits. Thus it was important to channel jovian or solar influences into the body of those afflicted with a melancholy or saturnine disposition. Since the principal vehicle of all such spirits was the air, therapeutic measures included not only the eating and drinking of appropriate food and libations but also the direction into the body of all favorable sensations borne on the air—scents emanating from appropriate sources; scenes, both natural and pictorial, which conveyed visually the proper "spirit"; and most importantly, sound in the form of apposite verse and music.[31] Following Walker's suggestion, we may

[30]D. P. Walker, *Spiritual and Demonic Magic from Ficino to Campanella* (London: The Warburg Institute, 1958); Frances A. Yates, *Giordano Bruno and the Hermetic Tradition* (Chicago: University of Chicago Press, 1964), esp. pp. 1–189. Also important for an understanding of the magical tradition of the Renaissance, particularly in German-speaking culture, are the two works of Will-Erich Peuckert, *Pansophie: Ein Versuch zur Geschichte der weissen und schwarzen Magie,* 2d ed. (Berlin: Erich Schmidt Verlag, 1956); and *Gabalia: Ein Versuch zur Geschichte der magia naturalis in 16. bis 18. Jahrhundert* (Berlin: Erich Schmidt Verlag, 1967), both of which explore the links with the German mystical tradition.

[31]The preceding account is based on Walker, *Spiritual and Demonic Magic,* pp. 3–72.

best picture one of Ficino's therapeutic sessions as a cross between a refined consciousness-raising experience, complete with *son et lumière* effects, and a religious ritual, with the participants decked out in appropriately fashioned vestments chanting sacred texts and music. [32] The cautious Ficino vigorously defended his magic as non-supernatural, dependent for its effects on the purely "natural" influences of the *spiritus.* He had some problems with his use of talismans, which sought to entrap the influences of beneficent spirits through the signs with which they were sealed, but his magic did not reach out beyond the sphere of the planets for fear of confronting the infernal demons lurking in the stars and realms beyond.

Ficino's magic found religious legitimation in the context of his Latin translation of the corpus of writings attributed to Hermes Trismegistus, which came into his possession, in their original Greek form, ca. 1463.[33] These texts in fact originated in the Hellenistic world of the first and second centuries A.D., as Isaac Casaubon first deduced in 1614.[34] They consist of a series of tracts in the form of revelations, which include an account of the Creation and recount a method of gnosis, both having intriguing parallels with the Christian tradition. They are also deeply imbued with Neoplatonic metaphysics and theology. This was what attracted Ficino. He believed these texts to be of preclassical origin, the genuine writings of an Egyptian sage and prophet, most probably a near contemporary of Moses, who in his account of the Creation gave a prominent position to the logos—the Word. Ficino believed he had found an independent Gentile prophet of Christ and Christianity, one whose revelations were imbued with Neoplatonic theological imagery. He cast about among the classical and patristic authors, finding principal support in Cicero and Lactantius, in an effort to construct a genealogy of this Gentile gnosis.[35] This genealogy, which never emerged in definitive form, ran roughly as follows: from the source, Hermes, an Egyptian prophet and sage at least contemporary with the Old Testament prophets; thence to Orpheus, who was the link with classical Greece (the Orphic hymns were an important element in Ficino's magical rites);[36] then to Pythagoras and

[32] Ibid., p. 30.

[33] The account of Ficino and Hermeticism which follows is based on Yates, *Bruno,* pp. 1–83.

[34] Ibid., pp. 170 and 398–402.

[35] On the various genealogies of sages related to Hermes see ibid., pp. 6–19; and D. P. Walker, *The Ancient Theology: Studies in Christian Platonism from the Fifteenth to the Eighteenth Centuries* (Ithaca: Cornell University Press, 1972), pp. 1–21.

[36] For a more detailed account of Orpheus's place in the *prisca theologia* see Walker, "Orpheus the Theologian," in *The Ancient Theology,* pp. 22–41; this

his followers; and finally to Plato himself. By this genealogy of Gentile prophets, Ficino attempted to restore Plato as the Christian philosopher *par excellence.*

Hermes was of course not entirely unknown to the Western world before Ficino's restoration of him. His name floated in and out of magical texts and especially alchemical ones. Indeed, one whole treatise of the Hermetic corpus existed in Latinized form as early as the ninth century, although it was falsely attributed to Apuleius of Madaura. [37] This work contained a description of certain Egyptian religious rites by which spirits where drawn down to animate statues of the gods. Such rituals were, of course, highly suspect in the Middle Ages, particularly in the light of Saint Augustine's specific condemnation of Egyptian religion and its statue magic.[38] But Ficino's rehabilitation of Hermes amongst prophets of Christianity and the sages of classical antiquity gave the magical arts so intimately associated with his name a new respectability.

To the Ficinian brand of Hermetic gnosis and magic, Giovanni Pico della Mirandola added a parallel Jewish gnosticism and magic, the Cabala.[39] Christian Cabalism was one of the more exotic offshoots of the Renaissance addiction to ancient letters, in this case Hebrew. The Cabala had its indigenous roots in Jewish mysticism and gnosticism. [40] Its powers derived from the characters of the sacred language itself, and operations were achieved by the manipulation of the Hebrew letters of the angelic and the divine names. In some of its forms numbers were attached to the alphabet, and the system could be worked out in terms of an elaborate numerology. Thus baldly stated, the Cabala would seem simply to have involved somewhat mechanical operations, but the whole was supported by and infused with a rich mystical and spiritual literature, which drew on some of the same Neoplatonic ideas repre-

---

chapter appeared in different form in *Journal of the Warburg and C. Institute* 16 (1953): 100–120. Walker discusses the rôle of the Orphic hymns in Ficino's magic in *Spiritual and Demonic Magic,* pp. 19–24.

[37]Yates, *Bruno,* pp. 3 and 9–10.

[38]St. Augustine's condemnation of the Egyptian magical religion was based on his reading of the Latin version of the *Asclepius* and is contained in the eighth book of *De Civitate Dei.* See Yates, *Bruno,* p. 9.

[39]There is no wholly satisfactory account of Cabalism in the Renaissance. Joseph L. Blau, *The Christian Interpretation of the Cabala in the Renaissance* (New York: Columbia University Press, 1944), is now somewhat dated. François Secret, *Le Zôhar chez les Kabbalistes chrétiens de la Renaissance* (Paris: Dunod, 1958), is good on diffusion of the movement but weak on interpretation. In what follows I have relied heavily on Yates, *Bruno,* pp. 84–116.

[40]On the Cabala in the Jewish tradition, see G. G. Scholem, *Major Trends in Jewish Mysticism* (Jerusalem: Schocken, 1941).

sented in the Hermetic gnosis. The Christian Cabalists naturally interpreted this Jewish mysticism in the light of their own theological presuppositions and used it to their own ends, including proselytization. From the perspective of Pico and later Christian devotees of the art, the Cabala represented a secret revelation given to Moses by God on Mount Sinai (in addition to the exoteric law inscribed on the tablets), which was subsequently transmitted orally to the elders.[41] The Christian Cabalists believed that this secret wisdom contained the vivifying interpretation of the "dead" letter of the law. Cabalism was both a gnosis and a magic, for it contained the creative powers of the language of God Himself—Hebrew. To Pico, the magic associated with the Cabala was much more powerful than Ficino's magic; it came directly from God, and in its operations on the angelic and divine names it sought to manipulate supernatural forces beyond the sphere of the stars. It was the magic of the Scriptures themselves.[42] Cabalistic magic for Pico took two main forms; the one subjective, the other transitive.[43] The subjective form was a process of mystical self-hypnosis, in which the soul of the operator, contemplating the appropriate characters and names, ascended gradually through the sphere of the fixed stars to the realm of the angels and the divine intelligences, culminating in an ecstatic vision of the Divinity. Transitive Cabalistic magic involved the production of magical effects on objects or other persons by means of the characters of the Hebrew language, or rather, by the power of the angels or Divinity invoked by these characters. This magic could be performed through talismans or numerical operations on the number values assigned to the alphabet. Such magic trod on dangerous theological ground, and Pico was indeed harried by the ecclesiastical authorities when he tried to present his magical theses in Rome in 1486.[44] It was quite explicitly a supernatural magic, and startling claims were made on behalf of it. Pico himself alleged that the divinity of Christ could be deduced from the operations of the Cabala on the divine names.[45] Furthermore, there was the suggestion, which Pico raised but firmly resisted, that Christ performed his miracles by Cabalism.[46] For Pico, the supernatural powers of Cabalistic gnosis and magic completed and perfected Ficino's Hermetic gnosis and its natural magic. Pico's aim was a new Christian synthesis of the Gentile and Hebrew traditions, not at the

[41]Yates, *Bruno,* p. 85.
[42]Ibid., p. 87.
[43]Ibid., pp. 93–116.
[44]Ibid., p. 112.
[45]Ibid., p. 105.
[46]Ibid., pp. 97 and 106.

philosophical and theological level but on gnostic and mystical grounds.

Nothing would appear to be more remote from the aesthetic splendors and scholarly profundities which characterized the Florence where Ficino and Pico wove their magic spells through the literary storehouse of Western culture than the tormented world of Paracelsus, with its endless journeys through the towns and villages of German-speaking Europe in search of an inner peace and a true vocation in the service of God. What a contrast between the Orphic hymnody of Ficino and the crabbed, often vulgar German prose with which Paracelsus flailed the scholarly establishment! Paracelsus scorned the men of the books; his yearning was for a direct confrontation with God and His Creation. And yet Paracelsus was no illiterate. His writings reveal a deep knowledge of the classical and scholastic medicine he so despised,[47] and still more of the gnostic and Neoplatonic currents of thought which he so admired and which are manifest throughout his works. In fact Marsilio Ficino is one of the very few people whom Paracelsus praised in his writings.[48] He modeled his treatise *De Vita Longa* on the Florentine's comparable major medical work.[49] Pagel has traced quite specific Ficinian influences in Paracelsus's medical theories and has even suggested that Paracelsus's whole life's work was an endeavor to implement Ficino's ideal of the priest-physician.[50] If indeed this was the case, it was a strange fulfillment of that vocation.

The work of Pagel has amply documented the gnostic, Neoplatonic, and magical strains in Paracelsus's thought.[51] However, the reforming physician's egocentricity, his bombastic style, and his peripatetic career make it extremely difficult to locate precisely the sources of these influences. What is certain is that they were refracted through a different cultural, social, and psychological experience. In the first place, Paracelsus was undoubtedly much more familiar with the gnostic, magical, and alchemical literature of the Middle Ages than the Florentines, who largely ignored this episode in their backward leap to antiquity.[52] Paracelsus himself tells us that he had immersed himself in

[47]Pagel, *Paracelsus*, p. 13.

[48]Ibid., p. 218.

[49]Ibid. The title of Ficino's work was *De Triplici Vita*, the first book of which dealt with preserving the health of scholars, the second with prolonging their life, and the third with astral influences on health. See Walker, *Spiritual and Demonic Magic*, p. 3.

[50]Pagel, *Paracelsus*, pp. 172–82 and 218–23.

[51]This is the major theme of Pagel's *Das Medizinische Weltbild des Paracelsus*. See also his "Paracelsus and the Neoplatonic and Gnostic Traditions," *Ambix* 8 (1960): 125–66. These themes are also treated extensively but less systematically in his English volume, *Paracelsus*.

[52]Pagel, *Paracelsus*, pp. 210–17 and passim. The same author treats these more systematically in his paper "Paracelsus: Traditionalism and Medieval

alchemy in his youth, and the medical-alchemical writings of such authors as Ramon Lull, Peter of Abano, Arnald of Villanova, and John of Rupescissa, as well as a selection of the host of anonymous medieval alchemical tracts, most probably figured prominently in his formation.[53] Other likely medieval sources of gnostic and Neoplatonic themes were the writings of Hildegard of Bingen (1098–1179) and the *Buch der Natur* of Konrad von Megenberg (1309–74).[54] In this context Hermes appeared, not as a great Gentile prophet of Christianity whose lineage linked him to the philosophers of classical antiquity, but as an obscure and ancient sage of alchemy. Also, Paracelsus's principal contact with the Renaissance magical tradition was through its German proponents. One of the few teachers acknowledged by Paracelsus who can be identified was Johannes Trithemius (1462–1516), the famed abbot of Sponheim whose speciality was the magic of seals and cryptography.[55] Agrippa von Nettesheim also spent some time under the tutelage of the abbot, and it is more than likely that Paracelsus was familiar with aspects of the occult magic of Agrippa.[56] In contrast to the Italians, these Germanic magi demonstrated more concern with the practical possibilities of magic than with the legitimacy of its bloodlines and the purity of its theology. It is from these contemporary German sources, as well as from some medieval references, that Paracelsus seems to have derived his rather imperfect knowledge of the Cabala.[57] The term occurs frequently in his writings, but he associated it with the

Sources," in *Medicine, Science and Culture: Historical Essays in Honor of Owsei Temkin,* ed. Lloyd G. Stevenson and Robert P. Multhauf (Baltimore: Johns Hopkins Press, 1968), pp. 51–75. This paper exemplifies well the peculiar difficulties associated with locating Paracelsus's sources.

[53]All of these possible influences have been discussed in Pagel, *Paracelsus:* for Lull, see pp. 241–47; for Peter of Abano, pp. 52, 214, 245, and 296; for Arnald of Villanova, pp. 248–57; for John of Rupescissa, pp. 263–66; and alchemical influences are discussed on pp. 266–73. In all of these cases, Pagel points out parallels with the thought of Paracelsus and thereby suggests influences. He also points out important divergencies. For Rupescissa, see also R. P. Multhauf, "John of Rupescissa and the Origin of Medical Chemistry," *Isis* 45 (1954): 359–67. For Paracelsus and medieval alchemy, see also W. Ganzenmüller, "Paracelsus und die Alchemie des Mittelalters," *Angewandte Chemie* 54 (1941): 427–31.

[54]On Hildegard, see Pagel, *Paracelsus,* pp. 210–12. Konrad von Megenberg is discussed in idem, "Paracelsus: Traditionalism and Medieval Sources," pp. 64–74.

[55]Pagel, *Paracelus,* p. 8. Trithemius's magic is discussed in Walker, *Spiritual and Demonic Magic,* pp. 85–90; see also Secret, *Le Zôhar chez les Kabbalistes chrétiens,* pp. 157–60.

[56]Pagel, *Paracelsus,* pp. 295–301. For a discussion of Agrippa's magic, see Walker, *Spiritual and Demonic Magic,* pp. 90–96; and Yates, *Bruno,* pp. 130–43. His career is discussed more fully in Charles G. Nauert, Jr., *Agrippa and the Crisis of Renaissance Thought* (Urbana: University of Illinois Press, 1965). In *Paracelsus,* p. 296, n. 301, Pagel notes that Paracelsus deprecated Agrippa's and Trithemius's knowledge of the occult.

[57]Pagel, *Paracelsus,* pp. 212–17 and 290–94.

magic of signs and seals (also known as the Gabalia)[58] and with a sort of telepathy by which imaginative impressions were transmitted through space—both topics which figure prominently in the writings of Trithemius. Certainly Paracelsus was no Hebraist, and in the *Philosophia Sagax* he dismisses Jewish magic as a corruption of Persian wisdom.[59] Thus, while it is legitimate to refer to Hermetic and Cabalistic strains in Paracelsus's thought, these are far removed from the full-blown scholarly syntheses of Ficino and Pico.

But even if Paracelsus had been thoroughly familiar with the Hermetic-Cabalistic syncretism of Pico, for social and psychological reasons, it would surely have had little effect on his attitudes or on his work. Paracelsus worked throughout his life amongst those who labored with their hands. It was from the peasant and the miner that he obtained much of his practical knowledge, and it was from their arts that he developed many of his concepts, images, and terms. He admired the artisans' intuitive wisdom and was sensitive to the "magic" of their techniques, which animated and made useful the gifts of God's creation.[60] In this company the scholarly esotericism of the Florentines would have appeared as pretentious vanity. Finally Paracelsus did not need the buttress of an unblemished gnostic heritage. His adopted motto, *alterius non sit qui suus esse potest,* speaks eloquently for his outlook.[61] To Paracelsus, his gnosis appeared to come directly from the light of nature and the light of grace.

Oswald Croll, in the *Admonitory Preface* to the *Basilica Chymica,* seeks to integrate Paracelsus's highly personal philosophy with the more historically conscious Hermetic-Cabalistic gnosis of Ficino and Pico. He takes Paracelsus's doctrine of the two lights and associates it intimately with the magic of Hermes and the Cabala. The synthesis is completed by the development of a singular theology of the Word, which sets Paracelsus's religious view of medicine in the context of a concisely articulated theological framework embracing Christian Hermeticism and Cabalism. In this formulation the seat of all wisdom remains Christ, and only Christians possess true knowledge.

58Peuckert defines Gabalia as the magic of the light of grace: ". . . der Magus ist der Mann, welcher dem Lichte der Natur folgt, und ist der Kenner ihrer Gesetze: der Gabalist wirkt durch das Licht der Gnaden durch Gott" (*Gabalia,* p. 20).

59*Philosophia Sagax,* in *Paracelsus, Sämtliche Werke,* 12:156–57. On this passage see also Secret, *Le Zôhar chez les Kabbalistes chrétiens,* pp. 295–96, who concludes that Paracelsus's Cabala is a form of astral and talismanic magic.

60As an adolescent Paracelsus had worked in the Fugger mines near Schwaz. See Pagel, *Paracelsus,* p. 13.

61As Pagel points out, (*Paracelsus,* p. 204), this motto must be taken with a grain of salt, but it is indicative of a state of mind.

All true Phylosophy should be grounded on the Scriptures and so return into God, that so the Regenerate Christians might reap and receive the full increase of that seed which among the Gentiles was choaked for want of the Sun, like that among the thorns: No Art can be perfected without Regeneration: True Phylosophy must be grounded on Christ the corner stone.[62]

This rebirth, however, was made available to Hermes and his followers before the Incarnation of Christ. Thus, the true philosophy of Christianity was foreshadowed by these ancient prophets and sages:

This Regeneration that holy man *Hermes* and others of clean hearts and godly lives before the *Word* was incarnate [had], being enlightned by the holy Spirit, though they concealed it among other Secrets, they knew it better then many of us who call our selves Christians and had rather seem to know God then love him.[63]

The "Divine Physick" of Paracelsus thus stretched back to antiquity through the Hermetic tradition, and Croll calls upon physicians to imitate the piety of these ancient doctors:

That which is the proper duty of an sincere, true and expert Physitian onely, who is more religiously and holily instructed, in the exhibition of their medicines they will follow the steps of pious and venerable antiquity; and imitate that most commendable and religious custome of the Hermetick Physitians, who always use to pray for a blessing upon their endeavours, striving to equall to those who will not take in hand the smallest matter without Divine Invocation. . . .[64]

For Croll, the source of all virtue in nature and hence of all natural medicaments was the Word. The Word in nature stemmed first from the creative "fiat" of God the Father. This was the Word that was manifest in the light of nature:

Whence we may observe that all things in the first Creation were produced out of the DIVINE NOTHING, or invisible Cabalisticall Poynt, into something, which God did in a moment; for his works cannot be delayed by time; All things proceeded out of the invisible Darknesse, and were called out to the visible Light by the WORD speaking, and the Spirit cherishing.[65]

But the Word of creation became the Word made flesh in Christ. Through the identity of the Word of creation and the Word incarnate Croll brings together the Paracelsian light of nature and light of grace

62*Admonitory Preface*, p. 52.
63Ibid., pp. 52–53.
64Ibid., pp. 156–57.
65Ibid., p. 70.

and focuses them on Christ. With Christ, the power of the invisible Word itself became manifest, and it was through this power that Christ cured during his earthly ministry:

> Physitians also have wrought great cures by the *Created Word,* or the incarnate Mercy: for all these things are done by the efficacy of the *Triune* and *Divine Word onely,* which healeth and preserveth all things, as we see in our Saviour['s] miracles, who when he restored the deafe and dumb (to whom the Pills and Syrups of all the Shops in the world could doe no good) he did it by *One Word,* and he is that *Word,* to wit, the increated Mercy of God. . . .[66]

Here Croll rushed in where Pico had feared to tread, for not only did Christ cure through the Word, but so did the prophets and Cabalists. This they accomplished through the powers of the letters of the name of God.

> These things wisely and rightly considered, we shall not wonder that Almighty God could (and can) make men whole by the *Prophets* and True *Cabalists* with a word onely. God is a living God, the NAME also of the living God is lively, and so the Letters of the living Name are also lively: God liveth for himselfe, his Name liveth because of him, the Letters live by reason of the Name; as God hath life in himselfe, so hath he given to his Name to have life in it selfe, and the Name also to the Letters.[67]

There is an unmistakable reference to the Cabalistic magic of the manipulation of the letters of the divine name. But Croll gives it a uniquely Christian twist, for the most powerful name is the name of Christ himself, Jesus.

> And the True Cabalists . . . he doth above Nature DEALLY or like God accomplish in a moment by firme confidence and strong faith, the very GATE of miracles in that *Only* Divine Name ISHUH in which all things are reckoned up and contained, that is he doth performe it in the WONDERFUL WORD by the Mind, Faith and Prayer, to wit, prayers made in Spirit and in Truth.[68]

[66]Ibid., p. 85.

[67]Ibid., p. 84.

[68]Ibid., pp. 87–88. Croll also considers the Cabala as a subjective mysticism, as in the following passage, where he describes the *mors osculi,* the last stage in the ascent to the ecstatic vision of God: "It is the intimate vision of GOD, and the intuitive knowledge of GOD, which also hapneth by the Light of Grace to the separate Soul even in this world, if any man set himself about it now, and be subject to God. Thus many holy men by vertue of the Deifick Spirit have tasted the First fruit of the Resurrection in this life, and have had a fore-taste of the Celestiall Country. I mean that Spirituall Death of the Saints (which the *Jewes* call the kisse of Death). . . " (*Admonitory Preface,* pp. 214–15). For Pico's discussion of this, see Yates, *Bruno,* p. 99.

By focusing the Cabalistic magic on the name of Christ himself, Croll could claim that this magic was nothing other than the fulfillment of St. Paul's injunction: "And whatsoever ye do in word or deed, do all in the name of the Lord Jesus, giving thanks to God and the Father by him" (Col. 3:17).[69]

By this tidy theology of the Word Croll links together basic Hermetic, Cabalistic, and Paracelsian themes. Hermeticism and Cabalism stand out ever more forcibly as gnoses of the Word. Hermes was a prophet of the logos–Christ. Cabalistic gnosis centers around the name of Jesus. Whereas for Pico, and also Agrippa, the magic of the Cabala tapped supernatural powers beyond the boundaries of the natural cosmos and lay end on as it were to natural magic, so Croll envisions both magics, natural and Cabalistic, as devolving from one source, the Word. In like manner Croll makes Paracelsus's two lights converge more sharply on the Word. The Word in nature is revealed in the light of nature; the Word incarnate is revealed in the light of grace. It is in the convergent focus of these two lights that medicine enjoys a special status amongst all men's arts: "Therefore among all Sciences and Faculties Physick is to be accounted the most excellent wherein the greatest wonders of God are miraculously seen. It taketh its rise from Theology or the Light of Grace, and endeth in the Light of Nature."[70]

69 *Admonitory Preface,* p. 157.
70 Ibid., p. 90.

CHAPTER II

THE ANATOMY OF THE WORLD

In this chapter we shall examine the Paracelsian world view which Croll presents in the *Admonitory Preface* to the *Basilica Chymica*. This view included an account of cosmogony, out of which evolved a coherent "theory" of physiological function, pathology, and therapy. It is, however, a theory based largely on gnostic insights. It entertains a vision of man as an integral part of the cosmos, operating in and through nature, not by virtue of the mastery of his reason over it but by the identification of his imagination with it. In the final section we shall return to the theological presuppositions which underlie Croll's synthesis and which give a unique character to his presentation of Paracelsianism.

"The happines of this present life," wrote Croll, "consisteth in the knowledge of Nature, therefore next to eternall things, in temporalls the chiefest thing is to find out the secrets of Nature."[1] Croll's theory of knowledge acquired by the light of nature and his cosmological views reflect closely those of Paracelsus. The fundamental doctrine in this world view was the macrocosm-microcosm analogy; the belief that there existed a correspondence between the powers and virtues operative in the universe (the macrocosm) and those operative within the organs of man himself (the microcosm). Man was viewed as a condensation of the whole universe. The significance of this doctrine for medical theory is at once apparent. The path to understanding the vital functions of man's body was through comprehension of the structure and operations of the great world. The invisible workings of

[1] *Philosophy Reformed and Improved in Four Profound Tractates. The I. Discovering the Great and Deep Mysteries of Nature: By that Learned Chymist and Physitian Osw: Crollius. The Other III. Discovering the Wonderfull Mysteries of the Creation, By Paracelsus: Being His Philosophy to the Athenians. Both made English by H. Pinnell, for the increase of Learning and true Knowledge* (London, 1657), p. 18, hereinafter to be cited as *Admonitory Preface.*

the body were revealed in the visible workings of the universe.[2] Croll expressed the parallel between the two worlds in terms of anatomy: "The outward World is a speculative Anotomy, wherein we may see, as in a glasse, the lesser World Man. . . ."[3] This anatomy of the world and of man, as Pagel and Rattansi have pointed out in their paper on Paracelsus and Vesalius, had little to do with conventional morbid anatomy.[4] Paracelsus and his followers, like Croll, were not interested so much in the gross structures of the human body as in the vital, and apparently invisible, forces which sustained life—and what could a dead body reveal about these? The true internal anatomy of living man was revealed only in the visible external anatomy of living nature.[5] Again in Croll's words:

[2]For an historical survey of the macrocosm-microcosm analogy in Western thought, see Rudolph Allers, "Microcosmus from Anaximandros to Paracelsus," *Traditio* 2 (1944): 319–407. The central position which this concept occupied in Paracelsus's thought is discussed in Walter Pagel, *Paracelsus: An Introduction to Philosophical Medicine in the Era of the Renaissance* (Basel and New York: S. Karger, 1958), pp. 50, 65–68, and passim. See also idem, *Das Medizinische Weltbild des Paracelsus: Seine Zusammenhänge mit Neuplatonismus und Gnosis* (Wiesbaden: Franz Steiner Verlag GmbH, 1962), esp. pp. 48–49; and also Kurt Goldammer, *Paracelsus: Natur und Offenbarung* (Hanover: Theodor Oppermann Verlag, 1953), pp. 43–46.

[3]*Admonitory Preface*, p. 24.

[4]Walter Pagel and Pyarali Rattansi, "Vesalius and Paracelsus," *Medical History* 8 (1964): 309–28, esp. pp. 310–14. I feel, however, that the authors place too much emphasis on the *chemical* anatomy, which, as I interpret it, is concerned mainly with pathological species and phenomena; see later discussion in this chapter.

[5]Paracelsus believed, in the context of the macrocosm-microcosm analogy, that by anatomizing the external visible heavens, the physician could penetrate to the living reality which lay behind the observer's eyes: "Wer wil dan ein artz sein, der den eussern himel nit erkent? dan im selbigen himel sind wir, und er ligt uns vor den augen, und der himel in uns ligt uns nit vor den augen, sonder hinder den augen; darumb so mögen wir ine nicht sehen. dan wer sicht durch die haut hinein? niemants. darumb vor den augen wachst der artz, und durch das vorder sicht er was hinder im ist, das ist, bei dem eussern sicht er das inner" (*Das Buch Paragranum Philippi Theophrasti von Hohenheim, beider arznei doctoris, in welchem die vier columnae, als nemlich philosophia, astronomia, alchimia und virtus, darauf et seine medicin fundiret, beschriben werden* [1530], in *Theophrast von Hohenheim, genannt Paracelsus, Sämtliche Werke. I. Abteilung: Medizinische, naturwissenschaftliche und philosophische Schriften*, ed. Karl Sudhoff, 14 vols. [Munich: R. Oldenbourg, 1922–33], 8: 97, hereinafter to be cited as *Paracelsus, Sämtliche Werke*). In another place Paracelsus compares anatomical dissection to a peasant looking at a psalter, who only sees the letters and understands nothing of their meaning; it is not enough: "nicht das gnugsam sei, so der cörper gesehen wird der menschen, item aufgeschnitten und aber besehen, item versoten und aber gesehen. das sehen is alein ein sehen wie ein baur, der ein psalter sicht, sicht alein die buchstaben; da ist weiter nichts mer von im zusagen" (*Labyrinthus medicorum errantium* [1537/38] in *Paracelsus, Sämtliche Werke,* 11:184). This passage is cited in Pagel and Rattansi,

> This is most evident from the Light of Nature which is nothing else but a divine Analogy of this visible world with the body of man; For whatsoever lyeth hid and unseen in Man, is made manifest in the visible Anotomy of the whole Universe, for the Microcosmicall Nature in Man is invisible and incomprehensible: Therefore in the visible and comprehensible Anotomy of the great World, all things are manifest as in their Parent: Heaven and Earth are Man's Parents, out of which Man last of all was created. He that knowes the parents and can Anotomize them, hath attained the true knowledge of their child Man, the most perfect creature in all his properties; because all things of the whole Universe meet in him as in the Centre, and the Anotomy of him in his Nature is the Anotomy of the whole world.[6]

The instruments of this anatomy of the great world were to be found in philosophy and astronomy, those two disciplines which Paracelsus characterized in his *Paragranum* as the first two pillars of medicine.[7] Paracelsian philosophy and astronomy had little to do with the traditional scholarly disciplines designated by these names. Paracelsian philosophy was concerned with the forces, "virtues," and activity of terrestrial phenomena revealed through "experience," and not with the material and formal characteristics of natural objects as deduced by ratiocination. Likewise, Paracelsian astronomy was not the construction or description of a mathematical model depicting the apparent movement of the heavenly bodies; it was a knowledge through "experience" of the powers and virtues of the celestial bodies and an understanding of the interplay of these astral influences with the workings of man's body.[8]

Nothing linked Paracelsian philosophy to the mainstream of Renaissance thought as strongly as the anthropocentrism enshrined in the macrocosm-microcosm analogy. The view of man as the focus and

"Vesalius and Paracelsus," p. 313. The *Paragranum* and *Labyrinthus* are two of Paracelsus's principal programmatic and polemical works.

[6]*Admonitory Preface*, pp. 24–25.

[7]The *Paragranum* was written in 1529/30, when Paracelsus was at the height of his creative powers. It contains some of the most virulent prose he directed at the scholarly and medical establishment. A good modern French translation by Bernard Gorceix now exists in *Paracelse: Oeuvres médicales* (Paris: Presses Universitaires de France, Collection Galien, 1968), pp. 29–100.

[8]Philosophy and astronomy were distinguished in terms of the elementary regions with which they dealt. Thus philosophy was concerned with the virtues of regions of earth and water, astronomy with the virtues of air and fire. "Nun is die astronomei hie der ander grund und begreift zwei teil des menschen, sein luft und sein feuer, zugleicherweis wie die philosophie begriffen hat auch zwen teil, die erden und das wasser" (*Paragranum*, p. 91). The other two pillars of medicine, as set out in the *Paragranum*, were alchemy and virtue. The former dealt with extraction of virtues from medicinal species and the latter with the corresponding virtues of the physician and his medicaments. These subjects are dealt with later in this chapter.

center of all creation, possessing all manner of natural, magical, and divine powers—given its most celebrated literary expression in Pico della Mirandola's *Oration on the Dignity of Man*[9]—found a cosmic rationale in the doctrine of the analogy of the great and the little world. Croll expatiated on the generation, dignity, and excellence of microcosmic man in a long section of his *Admonitory Preface.*[10] Man is the unique synthesis of the supernatural and the natural. He is set between time and eternity; above him lies the world of disembodied intelligences, the realm of the angels and God, in whose latter image he was created; and below him lies the sensible visible world of nature, out of which his temporal corruptible body was made. In his soul he bears the original archetype and pattern of creation which preexisted in the divine intelligence; in his body, the last visible product of God's creation, he bears the image of the great, sensible, and temporal world. "Man is made all things of God, and was therefore created last that by him the compleatnesse and perfection of all the Creatures might be signified."[11]

This goes at once to the heart of the Paracelsian theory of knowledge.[12] Man was no tabula rasa who acquired knowledge by the operation of the reasoning mind on the evidence of sensory observation. He possessed within himself all knowledge and all power of nature by virtue of his unique position in creation.[13] Not only was his soul endowed with a knowledge of the divine plan of nature, as in Plato's epistemology, but his body also united within itself all elements of that creation.[14] Man therefore had a knowledge of God and creation al-

[9]The *Oration* was to have been the opening statement in Pico's defense of his syncretic and magical theses at Rome in 1486; see Frances A. Yates, *Giordano Bruno and the Hermetic Tradition* (Chicago: University of Chicago Press, 1964), p. 86. The Latin text with Italian translation is in G. Pico della Mirandola, *De hominis dignitate, Heptaplus, De ente et uno, e scritti varii,* ed. E. Garin (Florence: Vallecchi Editore, 1942). An English translation will be found in E. Cassirer, P. O. Kristeller, and J. H. Randall, Jr., eds., *The Renaissance Philosophy of Man* (Chicago: University of Chicago Press, 1948), pp. 223–54.

[10]*Admonitory Preface,* pp. 53–75. Pico is referred to in marginalia on pp. 68–69.

[11]Ibid., p. 55.

[12]See Pagel, *Paracelsus,* pp. 50–65.

[13]"Because man hath the true and Reall possession of all things and Natures in himselfe, as also the speciall and perfect Image even of the Creator of all things; Therefore the knowledge of all things and natures, and of the Creator himselfe (wherein alone true Wisdome and Blessednesse consisteth) must take its rise from the knowledge of a mans selfe: So that Man, when he doth rightly understand himself, may in himselfe, as in a kind of Deified glasse, behold and understand all things" (*Admonitory Preface,* p. 48).

[14]"In respect of the Body or corruptible Nature he bears the Image of the great, sensible and temporall World; In respect of his soul or immor[tal] Nature, he bears the Image of the Archetype or originall copy and patterne of the world, that

ready within himself. This inner knowledge and the power inherent in it were awakened by "experience," that is, by a process of sympathetic attraction between the object of knowledge in the great world and its microcosmic representation within man himself. To be sure, this knowledge through experience involved a systematic empirical knowledge of natural objects in which the forms, characteristics, and virtues of those objects were properly differentiated and understood. But Paracelsian experience went beyond knowledge through empirical observation; it required a positive interaction between the particular object of knowing and its corresponding microcosmic equivalent in man. This form of knowing was much more akin to religious experience than it was to empirical knowledge, for it established an identity between the object and the knower and transformed the latter.[15] It should be borne in mind that the goals of Paracelsian understanding were the active powers and virtues of objects, not their external features, such as form and constitution, which were at best a sign of their virtue. Knowledge was thus never passive; it involved an interplay of powers and conferred power. Experience was justified in activity. For this reason the Paracelsians totally rejected the idea that wisdom could be gained from books.[16] True knowledge could only be gained in direct confrontation with nature. In this view they were at one with the spiritual enthusiasts who rejected the notion that religious enlightenment was to be found in the dead letter of Scripture. Natural experience gained through the light of nature paralleled mystical experience gained through the light of grace: both required a proper disposition, both transformed the recipient, and both conferred power.

is, of the immortall Wisdome of God himselfe" (ibid., p. 55). There is an important distinction here between man's knowledge of *nature,* which somehow resides in his *body,* and Wisdom, or divine knowledge, which is an attribute of the soul, where the *mens* resides. See further discussion below. Croll identified Wisdom with Plato's Ideas, present in the soul before it joined the body. Cf. ibid., p. 49.

[15]Paracelsus's own terms for experience are the German *Erfahrung* and the Latin *experientia.* His most detailed discussion of *experientia* occurs in the sixth chapter of the *Labyrinthus,* in *Paracelsus, Sämtliche Werke,* 11: 190–95. Here Paracelsus distinguished between *experimentum, scientia,* and *experientia.* An *experimentum* is simply an empirical knowledge of a property of a natural object, for example, the empirical knowledge that a particular herb has purgative properties. *Scientia,* on the other hand, is the inherent knowledge which the *natural object possesses,* which directs its function to its own proper end (that is, its *enteleche*). It is this *scientia* which the physician tries to seize through *experientia,* that is, through an intuitive understanding of the inner virtue of the object concerned, made possible by the macrocosm-microcosm relationship. For a discussion of this passage, see Pagel, *Paracelsus,* pp. 59–60; cf. also idem, "Religious Motives in the Medical Biology of the XVIIth Century," *Bull. Hist. Med.* 3 (1935): 98.

[16]"Physick is a favour given of God, whereof University books are not the Foundation . . ." (*Admonitory Preface,* p. 23, marginal note).

But not only was man the summation of creation—the microcosmos—he was also divine—the microtheos. Croll brings his passage on the dignity of man to a conclusion by stressing man's divine nature.[17] As God is the center of all things, all things returning to Him at the end of the world, so man, in the image of God, is the center of all creation. God set man as steward over His creation, and all creatures look to man as their guide and ruler: "On him all the Sphaeres bestow their beams, operations, reflections and influences, and on him all the Creatures poure their vertues and effects as upon a middle Point and Retinacle. . ."[18] This analogy between God and man was pointed up even more specifically by Croll. As God was one in essence and three in person, so man was one in person but three in essence. Man had three levels of existence, he encompassed three worlds: his mortal physical body, created by God from the slime of the earth (*è limo terrae*), which "he hath from the frame of the world and all things created therein"; his immortal soul, breathed into his body by God, which was the life of God dwelling in man; and uniting these two, the sidereal spirit, or astral body, which man derived from the stars.[19] The concept of the sidereal spirit, or astral body, was an important one in Neoplatonic theology.[20] Its function was to provide a bridge between the eternal world of ideas and the transitory world of the senses. The astral body was intermediate between pure spirit and corporeal substance: it was a pneuma of the finest corporeity, invisible and intangible. In Neoplatonic theology the astral body was conceived as a chariot which conducted the soul in its passage downwards through the stars to earth and which served to carry the soul back again through the stars in its ascent towards unification with the Divinity. The astral body, unlike the soul, was mortal.

In Paracelsian cosmosophy the astral body had a very significant rôle as the link between the supernatural and natural.[21] Its strictly theological rôle, however, was extended to make it the principal agent

[17]Ibid., pp. 55–75.

[18]Ibid., p. 56.

[19]Ibid., pp. 61–63.

[20]For a discussion of the astral body in this context, see *Proclus, The Elements of Theology,* revised text, with translation, introduction, and commentary by E. R. Dodds (Oxford: Clarendon Press, 1933), pp. 313–21.

[21]See Pagel, *Paracelsus,* pp. 117–21; *Das Medizinische Weltbild des Paracelsus,* pp. 54–62; and "Paracelsus and the Neoplatonic and Gnostic Traditions," *Ambix* 8 (1960): 127–32. For a discussion of the astral body in non-Paracelsian Renaissance medicine, with particular reference to Jean Fernel, see D. P. Walker, "The Astral Body in Renaissance Medicine," *J. of the Warburg and C. Inst.* 21 (1958): 119–33.

through which divine (vital) activity was transmitted and sustained in all of nature. Croll reiterates the analogy of the astral body with the chariot of the soul, "wherein the Intellectual soul and earthy Body like two Extreams are knit, glued and confederate[d] together, and in this third mean which partaketh of the other two they are coupled and united into one intire man."[22] It was through this medium that the soul was poured by God into the body of man. The astral body found its principal location in man's heart, from where it spread to all members of his body, having been joined to the spirits in the heart by means of natural heat and thereby diffused throughout the blood. Since this astral body originated in the stars, it kept the same circular course as that of the firmament. Croll presumably believed that the blood circulated in the body on the basis of such analogical reasoning.[23] Thus man had two bodies; his corruptible, visible body of flesh and blood and his invisible, insensible, astral body, which was derived from the stars. The astral body, as the source and seat of all vital activity in man, was the "true" body of man, "which moveth, guideth, and performeth all skilfull matters."[24] As such, it was the primary locus of Paracelsian physiological and medical theory.

The astral body, or sidereal spirit, however, was not unique to man in the sublunary world. Every living thing in nature–including metals and stones, which "grew" in the earth, as well as plants and animals–had its own astral body.[25] Like the astral body of man, it came from the stars and was transmitted through the air. Croll is at pains to stress that when he refers to the stars as the source of the sidereal spirit he is referring not to the visible bodies of the stars but to their invisible bodies. Indeed, he suggests an even higher source for the spirit: the invisible bodies are nothing less than the powers and virtues of the angels, which live upon the vision of God and are the created wisdom of God.[26] Thus the sidereal spirit had all the trappings of a Neoplatonic emanation. Its original source was the Divinity, and it was successively transmitted through the realm of the angels, or "intelligences," to the stars and from these to all living things in the sublunary realm. Viewed in this way, the sidereal spirit was a world soul. There is, however, something of a paradox involved in identifying the Paracelsian

[22] *Admonitory Preface*, p. 59.

[23] Cf. ibid., pp. 59–60. For a brief discussion of the circular symbolism in Croll, see Pagel, "Paracelsus and the Neoplatonic and Gnostic Traditions," p. 155; and *Das Medizinische Weltbild des Paracelsus*, p. 109.

[24] *Admonitory Preface*, p. 66.

[25] Pagel, *Paracelsus*, pp. 117–18.

[26] *Admonitory Preface*, p. 67.

sidereal spirit with such a Neoplatonic entity: whereas the concept of a world soul served to underline the inherent unity of the cosmos, in Paracelsian thought the sidereal spirit was at the same time the element which defined the uniqueness and specificity of all individual living objects.[27]

Every species of nature had its own sidereal spirit, or astrum, which, in the words of Croll, was "the secret Forger, from which every Formation, Figure and Colour of things proceedeth."[28] As the astral spirit penetrated matter, it became specified and gave form and function to the objects which it generated. This spirit is thus best comprehended, not as a continuous, homogeneous spirituous entity, but as the vehicle which contained and transmitted the totality of discrete specifying individual powers of nature. Croll gives as synonyms for the sidereal spirit the terms "spirit of the world," "nature," and Paracelsus's own word, *hylech.*[29] This identification of the sidereal spirit with nature eloquently confirms Pagel's definition of Paracelsus's own concept of nature as "the total of the acting Invisible, the spiritual forces that create form and are active in matter."[30] As the invisible astra descended through the cosmos they impregnated matter and generated visible growing things in their appropriate element—the elements being the Aristotelian quartet of fire, air, water, and earth, but here considered not as the universal material constituents of the universe but as the four principal cosmogonic regions, or matrices, of the sublunary world. Precisely how these "elements," or "wombs," provided the material matrix for the astra in Croll's schema will be considered below; for the present it is sufficient to recognize that all "growing" things were generated out of specific astra developing within an appropriate elemental matrix.

Croll, like Paracelsus, compares the astra to seeds.[31] The seed in any grain was the astrum, which, being cast into the earth, produced a visible body and in due time regenerated other seeds with further specific astra. The invisible astrum within the grain directed the assimilation of matter to produce the visible object as a species with its own characteristic form, quality, and virtue. It contained within itself the "knowledge" of the form and function of the particular object and directed its growth accordingly. Both Paracelsus and Croll compared

[27]Pagel, *Paracelsus,* pp. 104–5, comments on this inherent paradox in Paracelsus's philosophy.

[28]*Admonitory Preface,* p. 29.

[29]Ibid., p. 28; also see pp. 67–68 for other synonyms.

[30]Pagel, *Paracelsus,* pp. 224–25.

[31]*Admonitory Preface,* pp. 41–42 and 68.

the astrum to an inner *archeus*, or *vulcanus,* which forged and stamped the individual object with its specific form and virtue.[32] This it did, however, in a completely determined way, on a purely natural level, without the operation of reason; reason being an attribute exclusive to man in the creation.[33]

This emphasis on individual specificity as expressed in the doctrine of the astra was central to Paracelsian natural philosophy: it formed the core of his attack on the Aristotelian and Galenic view that the diversity of visible phenomena could be reduced to and deduced from the complexion of four qualities, or humors. Croll was forced to explain this unique individuality of the astra within the context of the macrocosm-microcosm analogy because there was a potential source of misunderstanding. Although the astra descended into the sublunary world from the stars, the virtues of the astra were not to be regarded as being determined by the stars:

> But when we say that all the form of things proceedeth from the astra's, it is not meant of the visible coales of Heaven, nor of the invisible body of the Astra's in the Firmament, but of every things own proper Astrum; so that the superior doth not power forth its vertues & hidden secrets into the inferiour spectificate Firmament, as the false Philosophers thinke that the stars of the Firmament do infuse virtue into herbs and trees; no in no wise: every growing and living thing carry its proper heaven and Astrum with it selfe, and in it selfe; the superiour stars in their course through the Zodiak excite and stir up the growth of inferiour things, they provide for them by dew, raine, seasons; but do not infuse the internall Astrum into things that grow, neither their smell, nor colour, nor forme, but all things proceed from the inner Astrum or secret forger, and not from without.[34]

Rather than a direct and strictly determined influence of the stars on terrestrial objects, there was, as this passage suggests, a much subtler harmony at work in the universe. Although the species of creation had their own unique individual lives, they cooperated with and were dependent on one another as members of the one family of nature:

> Neverthelesse the fruits of those Astra's or Caelestiall, Ayry, Earthy, Watry seeds doe indeavour and bend to one generall Good as Citizens of the same Anotomy: and therefore doe mutually cherish and succour one another by a sweet felloship and vicissitude of actions.[35]

[32]Ibid., p. 68. For Paracelsus's use of these terms, see Pagel, *Paracelsus,* pp. 105–6.

[33]*Admonitory Preface,* p. 68.

[34]Ibid., pp. 29–30.

[35]Ibid., p. 31.

This consideration of the astra affords us a more detailed insight into Paracelsian epistemology. As stressed above, man possessed within himself all knowledge of nature. The source and agency of this inner knowledge was the astral spirit. Just as the astral spirit emanated from the Godhead and descended through the cosmos carrying the individual astra which were brought to corporeal fruition in the wombs of the elements, so it was brought to focus again in the heart of man as the center of the microcosm. The astral body of man contained all the specifying astra of the universe, and at this level of his being man possessed inherently the knowledge of all the specifying powers in creation. He knew intimately the form, function, and virtue of every natural object because his astral body embraced the totality of the individual astra in nature. Croll expresses the situation in the following way, using Paracelsus's word *hylech* for the external astral spirit and the expression "Olimpick Spirit" for man's astral body:

> And as that Hylech in a particular manner containes all the Astra's in the great World, so also the internall Heaven of Man which is the Olimpick spirit, doth particularly comprehend all the Astra's. And thus the invisible Man is not onely all the Astra's, but is altogether one and the same thing with the Spirit of the world, as whitenesse is with snow . . . so also the visible corporall substances proceed from incorporall, spirituall (things) out of the Astra's, and are bodies of the Astra's, and remaine in the Astra's, one in the other.[36]

Later on in the *Admonitory Preface* Croll comments that if Aristotle and Galen had paid more attention to the "Olimpick Spirit" they would not have fallen into such errors in philosophy and medicine.[37]

The seat of man's knowledge of nature was thus not the mind but the heart, which was the focal point of man's astral spirit, from which it circulated to all the members of his body. The faculty associated with the astral spirit was not reason but the imagination.[38] The astra, which had no reason, generated things out of themselves by virtue of their "imagination." The imagination of man was identical with the sum of the imaginations of the astra, and man could, in imitation of nature, bring forth visible things by means of his arts, having first formed an image of them in his imagination. Thus man's knowledge of nature and his arts, which depend on that knowledge, arose from this communion of astra. Man, by virtue of his astral life,

[36]Ibid., p. 29.

[37]Ibid., p. 65.

[38]The imagination was the critical faculty in the Renaissance magical tradition. Through this faculty man gained a heightened sensitivity to natural and

himself became an *archeus* or *vulcanus* who could forge in his art all manner of wonderful objects in imitation of nature. As Croll expresses it in two passages from the *Admonitory Preface:*

> Astrum, Vulcanus and Archaeus are the same thing, and but one Spirit yet without Reason, & divers, as are the divers formes of severall things.[39]

and

> And whereas the imagination of Man is not one, but all the Astra's, it is as true that it produceth not onely one, but many operations. . . .[40]

The astral body was thus the seat of man's knowledge about nature and the source and inspiration of his crafts. Reason, on the other hand, was the source of man's knowledge about divine things and was the faculty of his soul. It was reason which elevated man above all other creatures and made him the microtheos. The astra, which determined the form, function, and activity of natural objects, operated with strict necessity; they had no reason to control or modify their effects. Man, on the other hand, possessed a rational soul, in which was enshrined an intellectual understanding of the divine plan (the archetype) of creation. Thus man could employ his reason to control the knowledge and power of nature, which he derived from the astra by virtue of his imagination, and direct them towards the ends which his soul dictated. All things, Croll asserts, naturally obey the soul and must of necessity move and work towards what the soul desires—"it makes all the powers of the world serve us, when by holinesse we draw vertue from him who is the true Archetype, and when we ascend to him, then every Creature must and will obey us and the whole Host of Heaven follow us."[41] His rational soul, illuminated by the light of grace, afforded man power over his imagination, illuminated by the light of nature. Man's reason also freed him from astrological determinism, which might, at first sight, seem a concomitant of the macrocosm-microcosm analogy.

> . . . the externall stars do neither incline nor necessitate Man, but Man rather inclines the Stars, and by his Magicall imagination infecteth them, and

divine truth, which could then be transmitted in artistic expression or in magical works. The imagination thus had subjective and transitive capacities. Frances Yates sees the changed status of the imagination as one of the principal distinctions between the mentality of the Middle Ages and that of the Renaissance; see *The Art of Memory* (Harmondsworth: Penguin Books, 1969), p. 226.

[39] *Admonitory Preface,* p. 68, marginal note.

[40] Ibid., p. 68.

[41] Ibid., p. 71.

causeth those deadly impressions; For we receive not our conditions, properties, and manner from the Ascendant, nor from the Constellation of the Planets, but from the hand of God through the breathing in of the breath of life; So that Mans Reason ought to rule the externall Stars.[42]

We must now return to the problem of how the astra become corporified in matrices of the elements to become visible, tangible, "living" objects. Thus far, we have a picture of creation proceeding from the logos, the divine Word of creation, which brought all the specific powers of nature into being. In a process perhaps best viewed as a cosmic precipitation, the powers of the logos descended through the cosmos in the form of astra, becoming increasingly less spiritual and increasingly more specific until, in the regions of the sublunary elements, they were materialized as the visible objects of nature. The Word became flesh first in the creation. At the point when the astra generated visible bodies in the elements, they were compared to seeds, a description intended to convey their status, in Pagel's words, as "threshold objects" between the Ideal and the Real, between Spirit and Matter.[43] Just how the elements provided the material substratum for the developing astra, however, presents formidable problems to the student of the Paracelsian tradition.

It is extremely difficult to deduce a coherent, still less a clear, theory of matter from Paracelsus's own corpus of writings.[44] This is perhaps not surprising, as one of the main themes of Paracelsus's philosophy was designed to counter the whole Aristotelian concept of cosmic change and diversity as a result of material and formal change.

[42]Ibid., p. 30. The statement in this quotation about man's imagination causing deadly impressions in the stars is a reference to Paracelsus's doctrine of the plague (see Pagel, *Paracelsus*, pp. 179–82). According to this doctrine, the plague originated with the evil passions and sinful imagination of man, which are "physically" transmitted to the planets and stars, to be revisited on man in the form of plague by the wrath of God; cf. *Admonitory Preface*, p. 69. Other psychophysical effects of the imagination are birth marks caused by the imaginative influences of the mother on the foetus and disease produced by the bite of a mad dog. The imagination thus could be the source of harmful effects. For further discussion of the pathological effects of the imagination in Paracelsus, see Pagel, "Paracelsus and the Neoplatonic and Gnostic Traditions," p. 149; and idem, "Religious Motives in the Medical Biology of the XVIIth Century," p. 103.

[43]Pagel, "Paracelsus and the Neoplatonic and Gnostic Traditions," p. 136.

[44]The most perceptive studies of Paracelsus's chemical theory are those by Ernst Darmstaedter, *Arznei und Alchemie*, Studies in the History of Medicine, ed. Karl Sudhoff and Henry E. Sigerist (Leipzig: Verlag Johann Ambrosius Barth, 1931); R. Hooykaas, "Die Elementenlehre des Paracelsus," *Janus* 39 (1935): 175–87; and T. P. Sherlock, "The Chemical Work of Paracelsus," *Ambix* 3 (1948): 33–63. Pagel, *Paracelsus*, pp. 82–104 and 273–78, draws heavily from these and other accounts.

The principal concerns of his philosophy were the spiritual forces operating within matter to produce change and the diversity of nature, in terms of the specificity of these virtues in nature. Matter per se did not interest Paracelsus, and in a sense it is even wrong to expect to find in his works a consistent theory of matter in either the classical or the modern sense. Nonetheless, Paracelsus did employ the four Aristotelian elemental categories of earth, air, fire, and water extensively in his writings. Again, there are significant differences in the status he gives these elements in his earlier writings as compared with that he gives them in his later works. In the most important of his early chemical texts, the *Archidoxis* (ca. 1526–27), he employs the four elements in a way perhaps closest to the meaning of Aristotle.[45] Here they are clearly intended as the material constituents of bodies. However, the properties of individual material bodies are not ascribed to a complexion of the pairs of qualities associated with each element; rather in every body there is a predestined element associated with a single quality which dominates and suppresses the other three. In his later works the four elements lose their significance as material constituents and take on the meaning of cosmic matrices, or "mothers," in which the semina of individual species are generated. To make an already complex situation more complicated, Paracelsus also introduced a new and original trio of categories, the *tria prima* of sulphur, salt, and mercury, into his discussion of the constitution of bodies.[46] Here again we find that the three principles are employed, not as material constituents, but as archetypes of specifying qualities whose status is not material but that of "threshold objects," or semina.

[45] *Neun Bücher Archidoxis* in *Paracelsus, Sämtliche Werke,* 3: 86–200. The *Archidoxis* was Paracelsus's principal "chemical" work and has been the major focus for those studying his contributions to chemistry. However, it is a comparatively early work in his corpus, and, as such, it draws more heavily on classical and medieval concepts than do his later writings. There is no mention in the *Archidoxis,* for instance, of Paracelsus's three principles (salt, sulphur, and mercury), generally regarded as his most original contribution to "chemical" theory. For a discussion of the *Archidoxis,* see Darmstaedter, *Arznei und Alchemie,* pp. 6–61; Hooykaas, "Die Elementenlehre des Paracelsus," pp. 175–180; and Sherlock, "The Chemical Work of Paracelsus," pp. 43–63. Sherlock believes that the *Archidoxis* represents Paracelsus's *system of chemistry,* but Hooykaas emphasizes the distinctiveness of the views presented there as compared with those in his later writings.

[46] The idea of the elements as matrices of species occurs throughout Paracelsus's works, most notably in the *Opus Paramirum* (1531), in *Paracelsus, Sämtliche Werke,* 9: 37–230 (cf. Hooykaas, "Die Elementenlehre des Paracelsus," pp. 180–84). The three principles make their appearance in writings composed in 1526, the same year as the *Archidoxis,* most notably *Das Buch de Mineralibus,* in *Paracelsus, Sämtliche Werke,* 3: 29–64 (cf. Pagel, *Paracelsus,* p. 103). For a discussion of possible antecedents of Paracelsus's three principles, see R. Hooykaas, "Chemical Trichotomy before Paracelsus?" *Arch. Int. d'Hist. des Sci.* 2 (1949): 1065–74.

All of this was as confusing to Paracelsus's commentators as it has been to subsequent historians of his ideas. One of the principal points of the ensuing debate has been the relationship between the four elements and the three principles. We catch an echo of the confusion that existed amongst the second generation of Paracelsian commentators in Croll's *Admonitory Preface.*[47] Croll initially sets out to clarify the interpretation of this point by one of Paracelsus's chief commentators and proponents, Peter Severinus (1542–1602).[48] After several pages, however, he abandons the attempt and sets out on his own interpretation. Croll at least succeeds in presenting a reasonably coherent scheme, and we shall follow him in this. But it should be borne in mind that this represents strictly Croll's viewpoint and is clearly a simplification of Paracelsus's own teaching. It is perhaps the most straightforward interpretation of material constitution of his generation of genuine Paracelsians.

Croll begins his discussion of the elements with a reference to the macrocosm-microcosm analogy.[49] The true philosophical physician must understand the frame of the external universe so that he might understand the inner constitution of the microcosm, man. For Croll the pillars of the frame of the great world are the four elements of earth, air, fire, and water. The elements, however, are not the material constituents of the universe; they are the cosmogonic regions, or wombs, in which the fruits of the astra are generated. The elemental wombs, as centers of vital activity, have a twofold nature: they are composed of spirit and body. The "true" element, the element of the element, is the spiritual part and is identified with the totality of the astra which are proper to that element, in the sense that they generate their fruit in that element.

> The true Elements with their proper Astra's are not visible or sensible, but as the Soul in the Body is insensible, so also are the Elements in their bodies. The body of the Element is a dead and dark thing; the Spirit is the life, and is divided into Astra's which out of themselves give their growth and fruit.[50]

Croll stresses that the dead and dark body of the elements

[47]The best discussion of the element theories among Paracelsus's followers remains R. Hooykaas, "Die Elementenlehre der Iatrochemiker," *Janus* 41 (1937): 1–28. Croll is discussed on pp. 15–16.

[48]Peter Severinus (or Soerenssen), a Danish royal physician, was one of the most influential spokesmen for Paracelsian medicine. His principal work was *Idea Medicinae Philosophicae Fundamenta* (Basel, 1571). Croll's discussion of Severinus's element theory is in the *Admonitory Preface*, pp. 33–37.

[49]*Admonitory Preface*, pp. 36–43.

[50]Ibid., p. 37.

would be powerless to generate growing things without the driving spirit of the astra, which in a sense know their own elemental womb and give rise to characteristic genera of visible species out of the body of the appropriate element. The fruits of the element fire are the stars and planets and all meteorological phenomena; those of the air are living creatures; those of water are metals, minerals, and salts; and those of earth are vegetables and plants. Thus visible, tangible objects are not formed from a *mixis* of all four elements, nor do they bear a characteristic quality associated with the element in which they are generated. Indeed Croll emphasizes that the astra and elements have no qualities in themselves—only their fruits have qualities—moreover, each individual element can give rise to species which have contrary qualities. Thus the fruits of fire include such contraries as snow and lightning; the fruits of earth include both "hot" and "cold" herbs. In addition, the individual fruits do not live out their lives constrained within the milieu of their maternal element, and indeed, they are most frequently manifest in an "alien" element. Minerals, salts, and metals are the fruit of water, yet they are dug up out of the earth; snow, rain, and dew are the fruits of fire, but they are manifest in the air and on the earth; corn grows in the earth but is reaped in the air. The element air has in fact a preeminent status in Croll's view, as it sustains the life of all created things, and none of the other three elements could bring forth their fruit without it.[51] What we have here is an aspect of that subtle harmony of nature operating at the elemental level. The elements cooperate with one another to sustain, nourish, and maintain a balance in nature appropriate to man's needs:

> . . . so the procreations of all the Elements doe voluntarily and earnestly bend toward Man-kind as to their desired limit, and by a liberal supply of moysture doe cherish all the parts of Nature; So also we see that by an imutable decree of Eternall Law it comes to passe and is so ordered that the Water doth not bring forth more then the Earth can bring up, the Aire cherish, and Fire consume.[52]

And what of the body of the elements? The bodies provide for the generation and regeneration of species; they ensure the continuity of the cyclical life of nature. The astrum of an individual grows to its mature corporeal form out of the body of its elemental womb and in due time generates seed in which new astra are encapsulated. When the body of the mature specimen withers and dies, these seeds are cast forth into the womb of the maternal element, where the body of the

[51] Ibid., p. 40.
[52] Ibid., pp. 39–40.

seed dies, and the liberated astrum generates a fresh body out of the body of the element. Without bodies there would be no continuity of species. Croll underlines this by asserting that the angels, unlike men, do not reproduce themselves, because they have no bodies.

> The Bodily growes out of the Spirituall, and abideth in it, and so the invisible vertues, Seeds and Astra's are propagated into many millions through the corporeall Visible body, as fire increaseth in wood or in convenient and fit matter, one Fire alwayes proceedeth from another. Angels cannot increase themselves because they want a body, but Man may because he hath a one.[53]

In Croll's scheme the body of the elements is composed of the three principles of salt, sulphur, and mercury, which are the matter out of which all things are created. Significantly enough, it is at this material level that Croll makes one of his very few appeals to strictly empirical proof. No body can be resolved into less than three constituents by chemical means, asserts Croll confidently, and in illustration he cites the example of burning wood, which is resolved into its three components: flame (sulphur); smoke (mercury); and ash (salt).[54] From this it is clear that the *tria prima* were not the chemical species we comprehend under the names sulphur, salt, and mercury but rather archetypes of qualities allegedly manifest in all bodies. Every material body had its own characteristic sulphur, salt, and mercury. Salt was the material archetype of solidity, color, and taste; sulphur was the material archetype of substance and transformation which tempered the coagulative properties of salt; and mercury was the archetype of volatility and fluidity which "by a diligent and constant supply of the vital and vegetative moysture doth cherish the two former. . . ."[55] But as with everything else in the Paracelsian world view, the empirically verified had a deeper, transcendental meaning, which was its ultimate justification. So it was with the *tria prima:* they were but a sign in the material world of the Trinity. "For the Holy Trinity when it spake that *Triune* word FIAT created all things *Triune,* as in a Spagiricall resolution is plainly to be seen."[56] As in the Godhead the Holy Spirit was the bond of love between Father and Son, and as in man the astral body united

[53]Ibid., p. 41.

[54]Ibid., p. 32.

[55]Ibid., p. 33.

[56]Ibid., p. 32. Paracelsus gave expression to this same view in *De Meteoris,* in *Paracelsus, Sämtliche Werke,* 13: 135. It seems certain that Croll had this passage in mind, as Paracelsus here gives as an example of the three principles white lead, red lead, and spirit of lead, an example Croll repeats on this same page; cf. Pagel, *Paracelsus,* p. 104, n. 271. For further discussion of the background to Paracelsus's

the soul to the body, so in matter sulphur bound the volatile mercury to the coagulative salt. The Mystery of the Trinity was thus signified at all levels of cosmic existence.

The prime matter of the universe having been created threefold, it was separated into the four elemental matrices, where it became the body of the elements. Under the directing influence of the individual astrum, the matter of the elements was assimilated to provide the body of the fruit of the astrum, so that each visible body had its own unique mercury, sulphur, and salt, the precise form and structure of which was determined by the astrum.

Pathology

The Paracelsian theory of disease was inextricably linked to the doctrine of the astra, the principles, and the elements.[57] Just as this doctrine was set in opposition to the Aristotelian concept of substantial change based on quality-endowed elements, so it also formed the basis of a pathological theory set in opposition to the Galenic theory of disease, which was based upon the complexion of the four humors—phlegm, blood, and yellow and black bile—with their associated qualities. According to the Galenic humoral pathology, there were no such ontological entities as diseases; there were but many manifestations of a distemper, arising from an imbalance of the four humors. Galenic therapy was directed at restoring the balance of the humors appropriate to the temperament of the individual concerned by a process of evacuation, such as bloodletting, diuresis, catharsis, etc., or by the administration of drugs, generally herbal in nature, which were thought to possess the appropriate degrees of hotness, dryness, coldness, and moistness. The Paracelsian theory argued on the basis of analogy of the great and the little world that diseases arose from the astra and the principles. Diseases were individual entities, each with their own unique pathological characteristics and localized seat in the body. There were as many diseases as there were astra in the universe; thus Croll could say that "no man knows the number of diseases but he that can tell the number of all things that grow."[58] The roots of disease were not the

principles, see Walter Pagel, "Paracelsus: Traditionalism and Medieval Sources," in *Medicine, Science and Culture,* ed. Lloyd G. Stevenson and Robert P. Multhauf, (Baltimore: Johns Hopkins Press, 1968), pp. 57–64.

[57]For a detailed survey of Paracelsus's pathology and other medical concepts, see Pagel, *Paracelsus,* pp. 126–202.

[58]*Admonitory Preface,* p. 44.

four qualities, which were only accidents or symptoms of the disease but not the disease itself.[59]

Disease was a result of the sin of Adam. As Croll expresses it, after the Fall the external corruptible part of man was set in opposition to his internal, invisible, whole self. As man was the microcosm, the Fall embraced the whole of nature; impurity, which was joined to the seeds of all things, became the source of all diseases.[60] As the sin of Adam resulted in man's banishment from the Garden and the dominance of his carnal nature over his spiritual self, so the various astra of creation, having lost the focus of their existence (man), were fated to live out their lives and multiply without reference to the overall plan and balance of nature. The astra, too, suffered dominance of their visible corruptible natures over their spiritual interiors. As a result of the Fall, the very diversity of creation and the imbalance between corporeal and spiritual which resulted became the source of disease.[61] The impurity of the astra arose from the dominance of their corporeal element over their spiritual element. In a normal state of health, there was a perfect balance among the invisible astra in man's body, which, as the microcosm, contained all the astra. Disease arose when this balance was disturbed by one of the astra generating its own fruit, which was manifest as a visible pathological symptom. "Man is a hidden world, because visible things in him are invisible, and when they are made visible then they are diseases, not health, as truly as he is the little world and not the great one."[62]

59For Croll's attack on the Galenic pathology and therapy of humors, see ibid., pp. 114–20.

60Ibid., pp. 95–96. In addition to the rupturing of the spiritual harmony of nature before the Fall, Croll suggests that God sowed specific "tinctures" of disease in nature as a curse on man, a suggestion which embodies the view of an avenging God, which was not really part of Paracelsus's own spirituality, although it is perfectly consistent with Croll's, as will become clear later in this chapter. The original Latin of Croll's statement on this point reads: "Post praevaricationem igitur & defectionem ab unitate ad alteritatem, maledictione Divinâ novae supervenerunt Tincturae . . . quarum mixtione in calamitosam sortem, una cum calamitoso mundi socio, transplanta est totius creaturae pulchritudo: Impuritas puris radicibus adjuncta, eaq[ue] fuit morborum praedestinatio" (*Basilica Chymica, continens philosophicam propria laborum experientia confirmatam descriptionem & usum remediorum chymicorum selectissimorum è lumine gratiae et naturae desumptorum. In fine libri additus est eiusdem Autoris Tractatus novus de Signaturis Rerum Internis* [Frankfort: 1609], p. 50).

61The notion that in its diversity nature carried the seeds of its own destruction was an important one in Paracelsus's world view. It carried the implication that man's diseases increased in severity and extent through historical time; cf. Pagel, *Paracelsus*, pp. 139–40.

62*Admonitory Preface*, p. 25.

Just as the fruit of the astra were generated out of the three corporeal principles of salt, sulphur, and mercury, so the pathological manifestations of disease, which likewise were fruits of the astra, were manifest as saline, sulphurous, and mercurial bodies. In the Paracelsian pathology all diseases were assigned to these generic categories: diseases characterized by skin eruptions were saline; inflammations and fevers were sulphurous; and those that proceeded from an excess of moisture were considered mercurial.[63] It was at this pathological level that the physician brought his knowledge of chemistry to bear. As in the macrocosm the astra generated their own specific fruit out of the three principles, so in the microcosm the astra generated their own specific fruits of disease. The diagnosis of a particular disease depended, therefore, upon the physician's skill in recognizing the parallel between the pathological fruit manifest in the body and its corresponding fruit in the great world. Thus the various species of saline diseases corresponded to the various types of salts found in the macrocosm. Recognition of these correspondences required a detailed knowledge of chemical species and properties.

An illustration of this chemical pathology is given in Croll's discussion of "mercurial" diseases. He states that mercury (and he clearly has in mind here the liquid metal we know by that name) can be exalted by heat to three degrees: firstly, by a mild digestive virtue (or heat), in which case it is distilled and gives rise to various types of apoplexies; secondly, it can be sublimed by the stronger heat generated by exercise, in which case it gives rise to frenzy and madness; and thirdly, it can be precipitated (that is calcined) by the heat of the stars, in which case the result is gout in the hands and feet.[64] Thus specific "mercurial" diseases were related to specific "chemical species" of mercury; the anatomy of disease became the anatomy of chemical species in the great world, or as Paracelsus called it, the *anatomia elementata.* There was a correspondence not only between disease species and chemical species but also between the seat of disease in the body and the location of the corresponding chemical species in the great world. Paracelsus referred to this anatomy of location as the *anatomia essata.*[65] A knowledge of both anatomies of disease was requisite for a sound pathology. In Croll's words: "It is not the Locall

[63]Ibid., pp. 121–23.

[64]Ibid., p. 122. Although all three principles could cause disease, mercury was, for Paracelsus, the prototype of a pathogenic agent, as it symbolized change; cf. Pagel, *Paracelsus*, p. 146.

[65]See Pagel, *Paracelsus*, pp. 136–38, for a discussion of the *anatomia elementata* and the *anatomia essata.* These two "anatomies" of chemical species were not as distinct for Paracelsus and his seventeenth-century followers as they

Anotomy of a man and dead corpses, but the Essentiated and Elemented Anotomy of the World and man that discovereth the disease and cure."[66]

We can now partially assess the rôle of chemical knowledge in Croll's (and Paracelsus's) paradigm of the anatomy of nature. Just as the first tools of the anatomy of nature—philosophy and astronomy—revealed the invisible workings of man's body (i.e., in its normal healthy condition) by analogy with the visible workings of heavenly and terrestrial bodies of the great world, so did the anatomy of chemical species reveal the pathological—hence visible—condition of man's diseased body. Philosophy and astronomy were the keys to understanding a healthy cosmos, in which the active powers worked in harmony. Chemistry was the key to understanding the diseased cosmos, which had fallen from grace with Adam and whose corporeal corruptible nature had gained ascendency over its interior spiritual nature.

Therapy

This disavowal of Galenic humoral pathology in favor of a chemical pathology necessarily involved a rejection of Galenic therapeutics. In the Galenic scheme, sickness being caused by an imbalance of humors, the object of therapy was to restore that balance either by direct or therapeutically induced evacuation of the excess humor or by "neutralization" employing pharmaceuticals of contrary qualities. To a Paracelsian such balancing of the humors did not go to the root of the disease; at best, the qualities were but secondary symptoms of the disease. Nature was not a complexion of contrary qualities but a repository of virtues (astra); and cures for diseased nature were accomplished not by restoring a balance of qualities but by restoring the harmony of virtues (astra). Croll compares the Galenic physician, who attempts to cure by adjustment of the complexion of qualities, with one who would extinguish the flame but leave the fire burning in the coals.[67] In some cases Croll urges therapeutic nihilism, arguing that the virtues of nature are best left alone to effect their own cures.[68] In cases where the intervention of the physician is required, he must strike at

might seem to us. Before the advent of systematic means of chemical analysis and identification, it was quite common to specify a mineral by the place where it was found.

[66] *Admonitory Preface*, p. 43.

[67] Ibid., p. 120.

[68] Ibid., pp. 117–18.

the root cause of the disease, namely, at the offending astrum. This does not imply that the astrum of disease is to be utterly destroyed. The astrum is not evil of itself; it simply has generated a visible corruptible fruit, and the object of therapy is to redeem it from this fallen state, by a revivification of its spiritual life.

Hence the object of Paracelsian therapy was to restore virtue to the diseased astrum. As every disease species in the microcosm had its corresponding species in the macrocosm, the therapeutic agent for the disease species was to be sought in the virtue of the corresponding species in the great world. Thus Croll can assert that the essential and elemental anatomy of man and the world reveals both the disease and the cure. The Galenic therapy of cure by contraries gave way to a Paracelsian therapy based on a homeopathic principle of like curing like: "Like things are said to be the Remedies of diseases, because they are of the same Anatomy of Nature, and because they have the like Signatures, Qualities and Roots."[69] A correct interpretation of the pathological species thereby led directly to a recognition of the corresponding therapeutic species. Since the totality of disease species corresponded to the totality of things that grew, it followed that remedies were to be found in all three kingdoms of nature.[70] However, since the fruit of the astra were generated out of the three principles of salt, sulphur, and mercury, there was a predisposition in Paracelsian therapeutics towards mineral remedies.[71]

Having diagnosed the therapeutic species corresponding to the diseased astral species in the body, the task of the physician is then to extract the "virtue" from the appropriate species in the great world and to administer it to the afflicted human body so that it might restore the spiritual life of the diseased astrum and recall it to its harmonious invisible function within the microcosm. To accomplish this task of medical redemption, the physician, as therapist, must first of all liberate the spiritual virtue of the macroscopic therapeutic species from its corruptible material matrix, "for," as Croll states, "our Remedies require preparations, separations and exaltations before they can impart their hidden and restrained vertues."[72] This is achieved by the alchemy of fire. Fire separates the pure from the impure, the spiritual from the

[69]Ibid., p. 121.

[70]Ibid., pp. 76–77.

[71]Ibid., pp. 100–103, Croll defends the Paracelsian use of mineral remedies. He argues among other things that stronger virtues were implanted in the solid fabric of metals and minerals and that nature could concentrate these virtues during their longer period of generation without distraction.

[72]Ibid., p. 93. On the redemption theme, cf. the following: " 'Tis necessary that the first life of hearbs and medicines should die that the second life by the

material, the virtue from the dross. Thus the physician-alchemist, in preparing his medicaments, must seek to extract the most active virtues from the appropriate therapeutic agents. By means of this alchemy the physician exalts his medicaments to a state in which they can be more readily assimilated by the body and directed to the precise seat of the disease. As alchemist, the physician performs a rôle similar to that of the *archeus* operating in the healthy body, for he separates the assimilable virtue out from the useless excrement. Indeed, the functions of the alchemist and the *archeus* are identical; man's alchemical art is but an exterior realization of his inner knowledge of the workings of nature, which is reposited in the astral body, of which the *archeus* is an element:

> For Chymistry . . . doth make manifest, not onely the true Simples, Wonders, Secrets, Mysteries, Vertues, Forces respecting health, but also in imitation of the Archaeon Ventricle or Naturall In-bred Chymist, it teacheth to segregate every mystery into its own reservacle, and to free the medicines from those scurvy raggs wherein they were wrapt up. . . .[73]

This medical alchemy is in fact only a part of a much broader enterprise which God has entrusted to mankind. Paracelsus set out the full dimensions of this alchemical activity in book 3 of his *Paragranum* and in book 5 of his *Labyrinthus medicorum errantium.*[74] The principal theme of these loci was that God had given to every product of nature a natural end which, in conformity with the anthropocentrism inherent in the macrocosm-microcosm doctrine, was defined in terms of man's needs. In addition, God had assigned to man the task of transforming, by means of alchemy, the raw products of nature into a state appropriate for man's utilization. Thus God had endowed man with crops, animals, minerals, and medicaments in all three realms of nature, but not necessarily in a condition to be immediately assimilated or utilized by man. Man had to garner them, segregate them, separate the pure from the impure, and bring them to perfection, usually employing fire at some stage. This was alchemy in its widest sense; and it made an alchemist not only of the physician but also of the farmer, the miller,

---

Chymists help may be attained through Putrefaction and Regeneration, wherein the Three First [i.e., principles] discover themselves with their hidden vertues, which are necessary for a Phisitian to know, for without Regeneration no hid Secret of Physick can be attained to, which is without all complexion of qualities" (ibid., p. 42).

73Ibid., p. 93.

74"Alchemia, der dritte grund medicinae," *Paragranum*, in *Paracelsus, Sämtliche Werke*, 8: 181–203; and "Das Fünft Capitel," *Labyrinthus*, in *Paracelsus, Sämtliche Werke*, 11: 186–90.

the baker, the stoker, the smelter, the smith—in short, of every craftsman who employed his skills in the preparation of nature's products for man's ends. This alchemy might involve more than one stage and more than one alchemist. To illustrate with Paracelsus's own favorite example, the alchemical preparation of bread involved the alchemist-farmer, who cultivated the wheat; the alchemist-miller, who separated the grain from the chaff; the alchemist-baker, who produced the loaf of bread in his alchemical oven. Nor did the alchemical process end there; for this exterior alchemy was completed in the stomach, where the inner alchemist brought the process to its end by concocting and perfecting the bread into flesh and blood. As Paracelsus expressed it in a key passage in the *Labyrinthus:*

> Thus, as follows from what I have said above, nature proceeds with us in God's creation in such a way that nothing is fully prepared in the form of ultimate matter. Instead all things are made as prime matter and subsequently the vulcanus acts upon it and makes it into ultimate matter through the art of alchemy. Then the archeus, the interior vulcanus, acts in the same way, for it knows how to circulate and prepare according to the components and the distribution [i.e., according to the *anatomia elementata* and the *anatomia essata*], as the art itself does with sublimation, distillation, reverberation, etc. For all of these arts are in men to some extent as they are in the outer alchemy, which prefigures them. Thus the vulcanus and the archeus separate [things] from each other. That is alchemy—that which brings to its end what has not come to its end; that which extracts lead from its ore and works it up to lead, this is part of [alchemy]. Thus there are alchemists of metals, and also alchemists who treat minerals, who make antimony into antimony, sulphur into sulphur, vitriol from vitriol, and salt into salt. Learn thus what alchemy is, and recognize that it alone is that which prepares the impure and makes it pure through fire.[75]

The socioreligious implications of Paracelsus's concept of alchemy were profound and revolutionary. Not only was the peasant-artisan elevated to the status of an alchemist, he was allotted a positive rôle in a great cosmic drama which was nothing less than the redemption of the world. Just as Christ redeemed man the microcosm, who had fallen from grace through the sin of Adam, so man in his turn would redeem the whole of nature, which had fallen with him, by separating the pure from the impure and refocusing the virtues and spiritual powers of nature on himself, the center of the great world. Thus the whole of nature would be redeemed—nature through man and

[75]*Labyrinthus*, in *Paracelsus, Sämtliche Werke*, 11: 188–89 (translation mine). Cf. Sherlock, "The Chemical Work of Paracelsus," p. 41, for a translation and discussion of this same passage. Sherlock was puzzled by the expression "nach den stücken und austeilung," which I take to refer to the *anatomia elementata* and *anatomia essata*, respectively.

man through Christ. This theology of the priesthood of the laborer was at the center of Paracelsus's social and religious challenge to his times.[76]

The soteriological dimension to Paracelsus's vision of man's alchemical crafts was set within the theological context of the doctrine of the two lights. For Paracelsus, the light of nature entered the world only after the expulsion of Adam from the Garden. It was the first expression of God's mercy for mankind after the Fall. By means of this light man learned to acquire the necessities of earthly life through tilling the earth, cultivating the soil, and extracting the virtues of nature which God had implanted at the Creation. Thus the mortal consequences of the Fall were tempered by this first act of divine mercy. A knowledge of the arts acquired through the natural light was not immediately conferred nor totally bestowed; it had to be garnered slowly, first by Adam and then by subsequent generations of men, accumulating experience and knowledge through the sweat of their brows. This light did, however, sustain man until that second great act of divine mercy, the incarnation of the Son of God.[77]

With Christ, the light of grace entered the world. This new light greatly outshone the light of nature. Whereas the natural light helped preserve man from mortal death, this new light of Christ redeemed man from eternal spiritual death. But for Paracelsus the light of grace did not displace or supplant the light of nature; rather it served to complement and fulfill it. Man continued to have need of the fruits of

[76]The intimate connection that existed in Paracelsus's mind between the works of man and Christ's act of redemption is brought out in the following passage from the *Philosophia Sagax* (translation mine): "But through man it has been revealed who He was, namely that Jesus of Nazareth was Christ, a king of the Jews. This has man's work revealed and made known as in the work of the cross, of [His] arrest, etc., [His] scourging, crowning and finally [His] death: man was the instrument for all of this. Thus in all things [man] is the revealer of all hidden things and moreover he is the man who is not seen, not even by himself. Just as Jesus was Christ, although the Jews saw Him only as Jesus but not as Christ—He was invisible. Thus it is the same with man himself, because he is hidden and no one sees what is in him except what is revealed in his work. Wherefore man should be continually at work so that he himself might learn to recognize what God has given him, and so that others might also see his works and thereby praise and glorify God from whom all things come, having been created and given" (in *Paracelsus, Sämtliche Werke*, 12: 59–60). The eschatological motifs in Paracelsus's thought are thoroughly explored by Kurt Goldammer in "Paracelsische Eschatologie: zum Verständnis der Anthropologie und Kosmologie Hohenheims," *Nova Acta Paracelsica* 5 (1948): 45–85 and 6 (1952): 68–102. The second part deals with the social and political implications of these views. See also Goldammer, *Paracelsus*, pp. 77–93; and Pagel, *Paracelsus*, p. 114.

[77]See particularly the passage in *Von den hinfallenden Siechtagen* (1530), in *Paracelsus, Sämtliche Werke*, 8: 290–92.

those who labored in the light of nature. Christ's enjoinder "Seek and ye shall find" applied equally to the fruits of nature and to the fruits of grace.[78] Christ's temporal acts of mercy, during His earthly mission, particularly His acts of healing, testified to the continuing significance of man's temporal works in and through nature. The powers of grace made accessible to man through Christ did not negate the powers of nature implanted by God the Father at the Creation. Christians in imitation of Christ's example had to employ both powers. But the powers inherent in both lights remained ultimately, for Paracelsus, gifts of God. Just as all men were not equally endowed with skill in their knowledge and crafts through nature, so not all men had equal powers conferred on them through the light of grace. Some were especially favored in the latter category and had the capacity of entering into the beatific life even in this world and of performing miracles in imitation of Christ—these were the saints—and others—the "natural saints," or "magi"—were especially favored in the light of nature and had great influence over natural powers and virtues.[79] These latter served God too when they used these powers according to God's will, that is, in conformity with the supernatural end of man as revealed in the new dispensation inaugurated by Christ. All Christians had to strive to walk in the paths of nature and of grace, each according to his own lights. The essence of the Christian life for Paracelsus was to make God's works visible in both lights.

Perhaps the most notable aspect of Paracelsus's theology was its Christocentrism, which was the counterpoise to the anthropocentrism of his natural gnosis. Christ, for Paracelsus, was preeminently the second Adam, and there is an especial emphasis in his writings on Christ's earthly mission, both His teaching and His works. The Fall and the Incarnation were the pivots of history. Adam brought the light of nature with him into the world, Christ the light of grace. Their respective powers defined the potential of human existence in the two periods of historical time. In all of this there is a remarkably positive outlook, a strong sentiment of the *felix culpa,* the Blessed fault of Adam which merited so great a redeemer as Christ.[80] Christ swept man up into His redemptive mission, for Christ lived on in the life of individual baptized

[78]"So wisset das wir von Christo ein mandat haben nach dem wir uns alle müssen richten und keren. nicht alein das sein praecepta und ler auf das ewig dienen sonder auch in das liecht der natur. und das selbige mandat is also: suchet so findet ir: das ist die kunst, die der mensch wissen sol, suchen und so findet ers" (*Philosophia Sagax,* in *Paracelsus, Sämtliche Werke,* 12: 150–51).

[79]See *Philosophia Sagax,* in *Paracelsus, Sämtliche Werke,* 12: 129–30.

[80]The phrase "O felix culpa, quae talem ac tantum meruit habere redemptorem!" occurs in the great *Exultet* prayer of the Roman liturgy on Easter Eve,

Christians and in the Eucharist. Christians were reborn into the life of Christ through Baptism and were obligated to continue His temporal and spiritual ministry. They were sustained in this life of Christ by partaking of His sacramental body in the Eucharist. At death man's elemental body, his Adamic flesh, would return to the dead matter of the elements whence it came; his astral body, which contained the invisible virtues of nature, would return to the realm of the stars and the angels; and his soul would return to the Godhead, where on the last day it would be joined to his glorified sacramental body, which was prefigured by participation in the Eucharist.[81]

When one turns to Croll's account of the relationship between God, nature, man, and his works, one is immediately conscious of a tenor in the underlying theology and spirituality which is completely different from that found in the writings of Paracelsus. The most striking feature of this spirituality is the stern and forbidding image it depicts of the relationship between God and man. In Croll one finds the connection between the light of grace and the light of nature defined in a different way from that of Paracelsus—a redefinition which greatly affects the spiritual value of man's works in and of themselves. Croll in effect recasts Paracelsus's philosophy in an essentially Calvinist theological and ethical framework by a skillful reordering of Paracelsian sentiment and expression combined with a singular interpretation of the Hermetic and Cabalistic tradition. What emerges is a Calvinist Paracelsianism set in an historical context which equates the magi and prophets of the *prisca* tradition in the old dispensation with the saints and elect of God in the new. The key to this transformation which Croll effects in the theological base of Paracelsianism lies in his (Croll's) Christology.

Christ, for Croll, was preeminently the logos—the Word Incarnate. There were for Croll two manifestations of the Word: The Word in nature, which stemmed from the cosmogonical utterance of God the Father, which reposited the specific virtues in nature; and the Word Incarnate—Christ, the Son of the Father and second person of the Trinity. The immediate effect of this almost exclusive identification of Christ with the logos was to bring to the forefront Christ's relationship with the Father at the expense of His relationship with man. Christ does not appear in Croll's Christology as the second Adam, who participates in the life and tribulations of man. The humanity and personality of Christ have, as it were, retreated into history. Christ

---

which accompanies the blessing of the Paschal candle, symbol of the risen Christ as light of the world.

[81]See Goldammer, *Paracelsus,* pp. 83–85; and idem, "Paracelsische Eschatologie," *Nova Acta Paracelsica* 5 (1948): 70–71 and 75.

ascended to the Father in historical time and only continues to operate in the world through the power of the Word, the Word of Scripture, and the efficacy of the Word made manifest by those who work in the name of Christ. It is from this twofold epiphany of the Word that the power of all healing is derived.

> Physick is two-fold in the Earth, Visible, which the Father hath created, & ought not to be administred before there can be a separation of the pure from the impure, Invisible, from the Son by the *Word,* and is but one. . . .[82]

All medicines operate through the power of the Word; those that are formed in nature are but a sign of the Word in nature.

> The Caelestiall Medicine onely, or the WORD of God (which is the Firmament of all Physick, without which no drug will doe good) is that which healeth all things, and by the efficacy of the WORD (in which lyeth hid, and from which proceedeth all force beyond any naturall actions) all Medicines become powerfull; As the bark is not the kernell, so hearbs are not the medicines, but a signe onely of the *Word* signified.[83]

We have noted above that Croll believed Christ to have cured through the power of the Word, a power which was imparted in turn to the prophets and true Cabalists, who cured through the divine name, Jesus. This power of curing through the Word also operated at an intermediate level between that employed by prophets and Cabalists and that employed by the physician working through nature, namely, through the magical effect of characters and signs impressed upon stones and amulets. This was the power of the Gamhaea, employed by the magi or celestial physicians, who linked the astral and terrestrial powers in their signs.[84]

When we turn to Croll's account of man's task of separating the pure from the impure by chemical means to gain the necessities of life, we find again that this activity has a different spiritual tenor and has been placed in a different theological context from that of Paracelsus. Croll presents it as follows:

> Nature therefore, as it is now, gives us nothing that is pure in the world, but hath mixed all things with many impurities, that as by the spur of necessity, it might often put us in mind that we should begin to learn the knowledge of Chymistry from our cradles, that so long as we are shut out of Paradice into the

82 *Admonitory Preface,* p. 83.
83 Ibid.
84 Ibid., pp. 87–90.

subburbs of this world, we ought to till and manure the EARTH , to wit, the whole frame of the world by admiration, searching into, and knowledge of both the Visible and Invisible (*Limus*) Earth, and that we should labour to get our bread, and other necessary things for this present life, as Natures Labourers, not lazily, but in the sweat of our browes, that by this means, by laying the Crosse upon us which we should bear with patience, it might stir up our industry in this LAND of LABOUR to attain the fruits of Terrene and Caelestiall Wisdome, least base and sluggish idlenesse make us wax leane and pine away, or (because we are more prone to all kind of sin and vice) by doing nothing we should learn to doe naughtily.[85]

There is little sense here of the *felix culpa.* Chemistry is not depicted as a blessing given in the light of nature to ameliorate the temporal consequences of the Fall; rather it is presented as a part of the divine retribution for the Fall. Man has to toil and learn the chemical arts if he is to ward off his physical demise; and behind this purely temporal demand lies a moral injunction to diligent labor lest he fall into the cardinal sin of sloth. The spiritual value of alchemical work is cast in a negative light, namely, as preserving man from idleness, which is the occasion of temptation and sin.[86] The classic equation of physical destitution with moral degeneracy, which became the rational of the seventeenth and eighteenth centuries' institutional response to the problems of poverty and vagabondage, lurks in the background of this passage.[87] How different from Paracelsus's own self-imposed vagabondage and identification with the outcasts of society! Croll has effectively removed chemistry and the manual arts from a soteriological context and placed them in the more familiar Calvinist terrain of ethics.

This passage in itself suggests that Croll viewed the relationship between the light of nature and the light of grace in a different way from Paracelsus. For the latter, the lights illuminated parallel paths which every man must follow according to his gifts. When Croll is examined more closely on this point, however, we find that the two lights are really not independent for him; in fact they are inextricably linked. We have already met Croll's insistence that "every true Theologer is a Phylosopher, and every true Phylosopher is a Theologer." The

[85]Ibid., pp. 96–97.

[86]The above passage of Croll should be compared with that of Paracelsus cited in fn. 77 and with a related one in the *Philosophia Sagax* (*Paracelsus, Sämtliche Werke,* 12: 241). In both of these places, Paracelsus stresses the *positive* aspect of God's beneficence in providing man with the fruits of nature and his arts through the light of nature so that he may enjoy to the full the benefits of these fruits. In the second passage, he stresses the positive value of man's labor in chasing away the devil: "dan solche arbeit unsers schweiss vertreibt den teufel und sein rott von uns, das keiner bleibet wo die arbeit ist."

[87]See Michel Foucault, *Madness and Civilization* (New York: Pantheon Books, 1965), pp. 56–57.

preamble to this conclusion makes the interconnection even clearer: "Without Phylosophy it is impossible to be absolutely godly; nor shall any man be ever able compleatly and Christianly to Phylosophize in either Light, who is not truely godly. . . ."[88] Godliness is the prerequisite for knowledge and power in both lights. Croll's theme is "Seek ye first the Kingdom of God, and his righteousness; and all things shall be added unto you" (Matt. 6:33). Primacy is given to the light of grace. Man has no independent path to knowledge and power through nature; the works of the old Adam simply serve to keep man out of mischief. But once man is reborn in the light of grace through Christ, natural knowledge and its fruits will come readily to him.

> Nor are we onely to know all Nature externally and internally, but we are also to make it our onely businesse, that according to the Fundamentall knowledge of the same by the supernall help of the *Light of Grace* we may together with Christ and all the Elect possesse that Eternall Life unto which God hath created us, this is true Theologicall Phylosophy: Wherefore the New Birth is first to be sought for, and then all other Naturall things will be added without much labour.[89]

The theological thrust of Croll's argument becomes clear in this passage, and it is patently Protestant in inspiration. Man's works in and of themselves have no value for salvation; they are only justified through Christ in the light of grace. Thus there can be no independent path to God in the light of nature, nor can man achieve any redemptive works in nature. To assert any such thing would plainly be contrary to the doctrine of justification by faith alone and would deny the sole efficacy of Christ's redeeming grace for salvation. Here we come back to Christ's relationship to the Father: it is only through Christ that we ascend to the Father. Croll takes this fundamental doctrine of Protestant theology and places his particular brand of Paracelsianism within its context; namely, that it is only through the Word of Christ in grace that we can gain access to the Word of the Father in nature.

Croll exhibits genuine Calvinist predilections in his categorical statement that God does not liberally dispense His grace to all but that "according to his good pleasure [he] may inspire whom he will, and deny it to whom he please."[90] Thus the wise men and magi in both lights become identified with God's elect. Croll celebrated Paracelsus above all as one of the elect, a member of the "*Intellectual* school of Pentecost, in which the Prophets, and Apostles, and all truly learned men walking in the Life and steps of Christ, have been taught and

88 *Admonitory Preface,* p. 135.
89 Ibid., pp. 131–32.
90 Ibid., p. 187.

learned without labour and toyle. . . ."[91] There is considerable irony in this, as Paracelsus had resigned himself to the belief that he was not especially favored in the light of grace and did not therefore belong in the company of the prophets and apostles.[92] He would no doubt have been flattered by Croll's inclusion of him amongst the saints, but his reaction to the assertion that his wisdom had been gained without labor and toil would most certainly have been unprintable.

Croll's Christian philosophers and physicians were thus an elect, divinely chosen and inspired. But they were not Calvin's elect, freely bearing witness to God's redeeming grace in their preaching of the Word and in their good works. They were an elect in an esoteric tradition, enjoined to preserve the secrets of divine wisdom and to prevent their abuse at the hands of the unworthy. They were in the tradition of the *prisca* theologians of old, who handed down the divine mysteries in a hidden manner. Croll conjures up all the wrath of the Divine Majesty to make his point:

> . . . so he that is an Adept and compleat Phylosopher, a keeper of Gods Secrets, and is conformed to the dignification of his work, after he by the Blessing of God hath happily labored like *Hermes* more then twenty years, fearing to offend the Divine Majesty, will be lesse afraid to die an hundred most cruell deaths, and indure all manner of miseries and punishment, rather then by any means, whether through wrath or what force soever shall be used, to publish to the wicked enemies of the Children of Art and Science, or to such as are unworthy of it, this greatest and richest Terrene Treasure, the Perfect Benefit or good gift of GOD, descending from the Father of Lights (as a Kingdom that will suffer no Compeer) from the King of Kings and Lord of Lords, that terrible and fearfull avenger of every unrighteous person, who hath intrusted him onely to keep it, which Secret would by evill men, with the ruine of the whole world, be abused to judgement or condemnation of an Eternal curse, because he knoweth most certainly that he shall be most grievously punished with the damnation of his soul, and losse of eternall salvation, by the Holy Trinity, and Christ the severe Judge of quick and dead, except he can give a good account of his Stewardship and the Talent committed to him, at the formidable and terrible Tribunal, before which we shall all stand, even of that Greatest and Eternall Judge (whose Terrible and Ineffable Majesty all mortall men ought to stand in fear of) who at the great Judgement day will examine our deserts, who will justifie the works of none that hath done evill, and deprive none that hath done well, of his reward.[93]

[91]Ibid., p. 136.

[92]Paracelsus of course believed that the Christian must strive to work in both lights, but the *Philosophia Sagax* marks his recognition of the fact that his special gifts were given in the light of nature. See Will-Erich Peuckert, *Gabalia: ein Versuch zur Geschichte der magia naturalis in 16. bis 18. Jahrhundert* (Berlin: Erich Schmidt Verlag, 1967), pp. 28–35.

[93]*Admonitory Preface*, pp. 190–91.

Croll's eclectic synthesis of Hermeticism and Calvinism is nowhere better illustrated than in this passage. The evocation of the Divine Majesty and the awesome depiction of God as the terrible but just judge of mankind is indicative of a deeply rooted Calvinist spirituality. Also noteworthy is the specific mention of Christ as participating in the judgment. Christ here is no second Adam, mediating for mankind. He has returned to the Godhead to assume His rôle as the "severe Judge of the quick and the dead." This echoes a recurrent theme in Calvin's theology, especially in his writings on the Eucharist, that Christ returned to heaven at His Ascension, His redeeming mission having been accomplished in historical time. Christ, therefore, cannot be brought down to earth again, as Roman doctrine of the Eucharist would suppose; rather He is to be sought in heaven through the agency of faith.[94] The overarching truth in both Calvin's and Croll's Christology was that Christ was God, not that Christ was once man. They both shared the fundamental position that man's spiritual ascent was via the Word to Christ and via Christ to the Father.

This identification of Christ with the Godhead rather than with man underscores Croll's insistence that man's works have no redemptive function in and of themselves. Man's works only acquire a positive spiritual value in light of their being favored and justified directly by Christ through grace, a favor which God had absolute freedom to bestow or withhold. The redemption of nature was thus reserved exclusively for God Himself, as Croll describes it in the following apocalyptic vision:

> When, I say, in that Terrible day he shall hold the exeltrees of the poles from turning about any longer, and the motion of the Elements shall cease, then all things shall run to ruine, and the heat of the Centre united to the heat of Sun shall set all corruption of the Elements on fire, when every evill and impure thing shall be cast like lead with the damned into Hell, where all things shall be tormented for ever, yet not consumed, with unquenchable brimston like molten glasse, continually burning and never wasting.
>
> And all that is of the Virtues and pure Truth and Nature of the Elements, which feare not the Fire of Heaven shall remaine like a pure, cleare, incorruptible and fixed Essence in a serene resplendent Chrystalin Earth, and be forever at rest with the happy saved Ones, carryed upward like an Eagle, or as a Smoak excited by the Fire: For when God shall change all things by making them new according to his will, and shall make them all like Christall, then the motions of the Supernaturall Nature shall abide in those things without corruption.[95]

[94]See *Short Treatise on the Lord's Supper,* in John Calvin, *Tracts and Treatises on the Doctrine and Worship of the Church,* trans. H. Beveridge, with historical notes by T. F. Torrance, 3 vols. (Grand Rapids: W. B. Eerdmans, 1958), 2: 164–98, esp. pp. 187 and 193.

[95]*Admonitory Preface,* p. 191.

The alchemical purification of nature is here exclusively attributed to a specific intervention of God acting "according to his will" on the last day. The redemption of nature is placed wholly within the eschatological setting of the end of the world. Man's works in nature, by contrast, have no soteriological function; they gain spiritual significance only insofar as they have been favored by grace, thus becoming particular testimonies or visible signs of God's Word operating in nature. All spiritual power and efficacy belongs to God alone; in no way can man constrain the miracle of grace. Oswald Croll has transformed the Paracelsian doctrine of the two lights in such a way that it has become compatible with Calvinist theology. This transformation was not without significance for the development of Paracelsianism in the seventeenth century.

Croll's preface to the *Basilica Chymica* is of course no orthodox expression of Calvinist theology, but the theological and spiritual framework in which his Paracelsianism is presented is clearly Calvinist in inspiration. The outlines of the structure are clear: the emphasis on God's majesty and justice; the primacy of the Word as the source of all wisdom and knowledge of God; Christ as the sole redeemer and mediator of mankind; justification achieved solely by faith; the doctrine of election; the futility of man's works, except that they be justified by grace. The parallels are clear; but the substance of Croll's message in the *Basilica Chymica* is obviously very different from that of Calvin in the *Institutes.* The essence of that difference lies in the fact that their respective world views were rooted in two distinctive scholarly traditions that gained new emphasis in the Renaissance. The Paracelsian Croll belonged to the Hermetic, Neoplatonic, magical strain of the Renaissance, which sought to recover the secret wisdom of antiquity; Calvin belonged to the humanist, philological, and exegetical strain, which sought to recover the pristine message of antiquity—in Calvin's case preeminently the pristine message of the Scriptures—by means of critical literary scholarship. The contrast between these opposing approaches to the intellectual inheritance of Western Europe which so gripped the sixteenth century stands out with particular clarity in the case of Croll and Calvin because of the very parallels in the structures of their thought. It can be encapsulated in one simple but profound distinction—the signification of the Word. For Croll the Word in the Scriptures and the Word in nature were signs whose true meaning was hidden beneath the surface and whose signification was only vouchsafed to those who were illuminated from within by the light of grace and of nature. On those favored by this light devolved the magic powers of the Word, the Cabalistic magic of the Word in the Scriptures and the natural magic of the Word in creation. For Calvin, on the other hand, the Word in the Scriptures contained the message of man's spiritual destiny, but not in any hidden or secret manner. He applied all of his expository and exegetical skills of language and criticism to lay this message bare. To those chosen in the light of grace came not any magical power but simply the ability to recognize and give witness to God as Father and Savior of mankind.

These two viewpoints were not mutually exclusive, however. Calvin did entertain the idea of the Word as sign in one very important part of his theology, namely, in his doctrine of the sacraments. While recognizing the priority of the Word in the Scriptures, Calvin viewed the two sacraments of Baptism and the Eucharist as special signs of the Word instituted by Christ to confirm in a particular way the redeeming message of the Scriptures. As he puts it succinctly in his *Short Treatise on the Lord's Supper,* with reference to the Eucharist: "For seeing we are so weak that we cannot receive him [Christ] with true heartfelt trust, when he is presented to us by simple doctrine and preaching, the Father of mercy, disdaining not to condescend in this matter to our infirmity, has been pleased to add to his word a visible sign, by which he might represent the substance of his promises, to confirm and fortify us by delivering us from all doubt and uncertainty."[96] The elements of the signs—the water of Baptism, the bread and wine of the Eucharist—convey explicitly and precisely the spiritual function and reality which is intended in the sacraments. Wishing to steer a narrow course between the doctrine of consubstantiation of Luther and the theories of Zwingli and Oecompladius, which represented the sacraments as symbolic commemorations of Christ's acts, Calvin draws on Augustine's presentation of the sacraments as nonverbal signs of the word. He is insistent that the sign is more than a mere symbol, as the spiritual reality is combined with sign; but at the same time there is no substantial change in the elements of the sacrament, since then there would be no point to the specific forms of the sacraments—they would have lost their import as signs.[97] Calvin further underlines the connection between the sacramental signs and the Word by insisting that their administration be accompanied by the preaching of the Scriptures.[98] Calvin strictly limits the manifestation of the Word

[96]*Short Treatise on the Lord's Supper,* 2: 166.

[97]See John Calvin, *Institution de la Religion Chrétienne,* critical edition with introduction, notes, and variants, ed. Jean-Daniel Benoit, 5 vols. (Paris: Librairie Philosophique J. Vrin, 1957–63), 4: 289–317. For a more succinct statement of Calvin's doctrine of the sacraments as nonverbal signs, see *Short Treatise on the Lord's Supper,* 2: 186.

[98]*Short Treatise on the Lord's Supper,* 2: 48–49. Calvin's doctrine of the

as sign to the two sacraments of Baptism and the Lord's Supper; and their administration is restricted to those properly authorized to preach the Word. The sacramental signs are the preserve of the organized visible Church and in a sense are themselves a further sign of the unity, faith, and confession of the true Church.

Calvin was prepared to admit that God had given many more specific signs of His Word in the Old Testament, which might properly be called sacraments. These were specific revelations of testimonies of God's paternal beneficence to His chosen people, necessary to confirm their faith in the covenant. They took the form of miracles and signs in nature. This latter category included the tree of knowledge of good and evil in the Garden of Eden, the rainbow manifest to Noah during the flood, and the burning bush given as a sign to Abraham. But in the new covenant between Christ and His Church such special revelations were unnecessary, and the sacramental signs were limited to two.[99]

It is here that Croll's universal interpretation of the Word as sign leads him furthest from Calvin. From Croll's standpoint all the specific powers of nature as signs of the Word become potentially sacramental to the elect.[100] The difference between Croll and Calvin is perhaps best brought out in their interpretations of the article of the Lord's Prayer "Give us this day our daily bread." For the latter, this petition refers solely to man's daily temporal needs and is in no way associated with the spiritual nourishment of the Eucharist.[101] Croll on the other hand gives the following interpretation:

> This WORD of *God,* the First begotten of every Creature, is truly our *Dayly Bread* for which our Saviour commanded us to pray; it is the supercaelestiall Mummy, the supernaturall Balsome comforting poor Mortalls more than Mans own

---

sacraments as nonverbal signs is wholly Augustinian. Primacy is given to the preaching of the Word in the Scriptures, that is, to the literal Word; the sacraments are signs confirming the Word. Cf. the following in the "Mutual Consent of the Churches of Zurich and Geneva as to the Sacraments," in Calvin, *Tracts and Treatises,* 2: 214: "For although they [i.e., the sacraments] signify nothing else than is announced to us by the word itself, yet it is a great matter, *first,* that there is submitted to our eye a kind of living image which makes a deeper impression on the senses . . . and *secondly,* that what the mouth of God had announced is, as it were, confirmed and ratified by seals." For an interesting discussion of St. Augustine's linguistic epistemology, which had so much influence on Calvin, see, Marcia L. Colish, *The Mirror of Language: A Study in the Medieval Theory of Knowledge* (New Haven and London: Yale University Press, 1968), pp. 8–81.

99 Calvin, *Institution,* 4: 307–11.

100 For Croll, in contradistinction to Augustine and Calvin, the nonverbal sign has primacy. Literal words, both in the Scriptures and in men's teachings about nature, are only imperfect statements of the signified Word.

101 Calvin, *Institution,* 3: 389–92.

Mummy or naturall Balsom. The vertue in Bread is the Blessing of God, yea God himselfe. . . .[102]

Croll's world of the Word signified throughout nature makes the whole of nature sacramental. As for Calvin, the efficacy and power of the sacramental signs for Croll were only accessible to the elect in grace. But Croll's sacraments and his elect were not contained within the body of a visible Church; his enthusiastic and mystical form of Christianity only recognized an inner mystical union of the elect, which cut across the denominational boundaries of institutional Christianity. Thus he could include amonst his elect the heterodox Catholic Paracelsus and the heterodox Lutheran Valentin Weigel. Along with other expositions of a mystical Christianity in the enthusiastic tradition of the sixteenth and seventeenth centuries, there is a strong irenicism in Croll's preface, but it is an irenicism which has very strong affinities with doctrinal positions in orthodox Calvinism.

This irenical appcal of Croll from the doctrinal base of Calvinism is in many ways the most intriguing part of the work, for it reveals what the *Basilica Chymica* really was: not just a textbook of Paracelsian chemiatric pharmacy, but a political document as well. The *Admonitory Preface* of the *Basilica Chymica* was dedicated to Croll's patron, the Calvinist champion of central Europe, Prince Christian of Anhalt, for whom Croll also worked as a political agent in Prague toward the establishment of an Evangelical Union of Protestant Princes against the House of Hapsburg. Croll's diplomatic rôle in these negotiations was to win over to Anhalt's side the support of the influential Bohemian Prince Peter Vok of Rozmberk, to whom the third part of the *Basilica Chymica*, the treatise on signatures, is dedicated.[103] Vok was an important but refractory element in the Bohemian phase of Anhalt's policy. A mysterious and somewhat tragic figure–his wife was mad and he had no heirs–Vok wavered in his doctrinal position throughout his life. He early showed leanings towards Lutheranism and subsequently towards the enthusiastic Czech Brethren. But he was also one of the principal collectors of the manuscripts of Paracelsus, most notably his theological writings. Croll praises Vok in the dedication to the treatise on signatures for his devotion to the true philosophy of nature. The *Basilica Chymica* thus surely reveals the ideological and religious grounds on which Anhalt, through Croll, made his appeal to

[102] *Admonitory Preface*, p. 86.

[103] For a discussion of Peter Vok and his relationship with Croll and Anhalt, see R. J. W. Evans, *Rudolf II and His World: A Study in Intellectual History, 1576–1612* (Oxford: Clarendon Press, 1973), pp. 140–43, and Gerald Schröder, "Oswald Croll," *Pharm. Ind.* 21 (1959): 407–8.

one of the magnates of Bohemia, an appeal which countenanced Vok's Paracelsian interests but which did not compromise Anhalt's Calvinism.[104]

[104]Schröder also suggests that Anhalt had designs on the heirless Vok's estate. His information is apparently based on diplomatic documents of Anhalt, for which, unfortunately, he does not give specific sources. The story had a curious ending. Croll had infiltrated Rudolf's court on the basis of his Hermetic and Paracelsian interests. The emperor, Evans relates (p. 142), made great efforts after Croll's death to acquire Croll's effects, which supposedly contained much occult and secret lore. The unexpected secret which they revealed was Croll's duplicity (see Schröder, "Oswald Croll," p. 409).

CHAPTER III

PRAXIS AND THE WORD

One of the most far-reaching proposals of the Paracelsian reform of medicine was the call for a renewed alliance of craft and knowledge in the medical art. Croll specifically asserts in the *Admonitory Preface* that "a Physitian therefore should have both the Theory and Practice, he must both know and prepare his medicines. . . ."[1] Such a proposal struck at the time-honored division of medical practitioners into physicians, who diagnosed and prescribed on the basis of their academically acquired knowledge of medical theory; apothecaries, who fulfilled the physicians' prescriptions in their capacity as retail tradesmen; and surgeons, whose responsibilities and functions were dependent on manual skills in the healing and treatment of external lesions, the setting of bones, and the performance of certain internal operations—skills acquired, like those of the apothecary, in the context of a guild-apprenticeship system. The Paracelsian reform envisaged the collapse of the educational and social barriers which delimited the medical profession in this way. With reference to the apothecaries, Croll writes:

> The choice also of the Medicines must alway be considered, and their preparations and compositions made by the Physitian himselfe, and not carelesly left to others. He is truly a genuine Physitian who can tell how (not onely by Reason, as mear Rationall Physitians doe, but) by their own hand to prepare the medicaments. . . .[2]

And again in reference to the surgeons:

[1]*Philosophy Reformed and Improved in Four Profound Tractates. The I. Discovering the Great and Deep Mysteries of Nature: By that Learned Chymist and Physitian Osw: Crollius. The Other III. Discovering the Wonderfull Mysteries of the Creation, By Paracelsus: Being His Philosophy to the Athenians. Both made English by H. Pinnell, for the increase of Learning and true Knowledge* (London, 1657), p. 152 (hereinafter to be cited as *Admonitory Preface*).

[2]Ibid., pp. 151–52.

> . . . it is necessary that every Surgeon should be a Physitian, and every Physitian a Chyrurgion, that there may be a sound Bridgroom for a sound Bride.[3]

Such a view stemmed in part from Paracelsus's celebration of the wisdom, skills, and virtue of the craftsman and artisan over those of the scholar and was to some extent motivated by social ideology. But it ran deeper than that: it was intimately linked to the Paracelsian conception of the nature of knowledge itself and to the religious significance of its responsibilities.

Knowledge for the Paracelsians was not a dialogue between experience and intellect moderated by reason: it arose from the action of the light of nature on the imagination to produce a comprehension of the powers and activities of the virtues of nature. Man derived his arts and his crafts from this light, for these were but externalized and controlled manifestations of the inherent powers of nature. Reason served to direct the knowledge of such power in conformity with the overall divine plan for nature and for man. Such knowledge was a unique gift of God, granted to each individual according to his own lights, so that, as Croll states, "in this study, no man is further to be believed, then as everyone findeth by his own proper experience."[4] The intuitive and experiential basis of this knowledge severely limited, if it did not entirely preclude, an objective description of natural phenomena. Man could not separate himself in time and space from natural events. Nature was within him as well as without him, and all was encompassed within God. As such, man was inextricably caught up in the pulse of cosmic events stretching from the Divinity to the lowest of the elements—he was a *minister to,* and not a *master of*, nature, as Croll puts it.[5] Given such a condition, man could not possibly articulate a theory of nature which determined practice according to the dictates of reason. Paracelsianism, in Croll's formulation, did not set down so much the principles of a natural philosophy; it was more like a guide to

[3]Ibid., p. 151. The Paracelsians were not alone in their call for a renewed alliance of craft and theory in medicine during the sixteenth century. Vesalius, in the preface to his *De humani corporis fabrica,* argued the same point. His appeal, however, was couched in characteristically humanist terms, calling for a restoration of the integrity of Greek medicine, in which all the sects recognized the importance of the physician using his hands. See the English translation of the preface to the *Fabrica* in C. D. O'Malley, *Andreas Vesalius of Brussels, 1514–1564* (Berkeley and Los Angeles: University of California Press, 1964), pp. 317–24. The Paracelsians' view stemmed more from the religious base of their ideology, but they did make frequent appeal to Hippocratic precedent for the alliance of medicine and surgery. Cf. Croll, *Admonitory Preface,* p. 151. I am grateful to Dr. Owsei Temkin for drawing my attention to this point.

[4]*Admonitory Preface,* p. 5.

[5]Ibid., p. 128.

a natural gnosis. Still less could the knowledge of the powers of nature be transmitted by words: it must be expressed in action. The Paracelsian physician always sought justification in terms of the effectiveness of his cures. Such effectiveness was viewed, not as the empirical validation of theory, but as the shining witness of the authenticity of experience. Furthermore, since experience was unique to each individual and dependent on his own spiritual disposition, action proceeding from that experience could not be delegated. The Paracelsian physician must be his own apothecary and his own surgeon. In this spirit Croll criticizes the bedside antics of the academically trained physician:

> . . . yet these Rationall Physitians when they come to the sick mans bed know not what to doe or which way to turne themselves, but stand wondering and as men amazed, speaking smoothly, and giving their patient a parcell of good words onely, being able to doe nothing toward his recovery, because they can prepare their medicines but onely with their Reason, not at all with their hand.[6]

But the relationship of experience and practice in Paracelsian medicine also had its deeply religious context. The religion of Paracelsus and his followers was above all a religion of works. Medicine derived its special significance from its association with Christ. Paracelsus saw medicine in the context of Christ's earthly mission. He saw two aspects of Christ's ministry as inextricably linked—the preaching of the Word and the healing of the sick. Both ministries were directed at the center of God's creation, man: the ministry of healing delivered man from natural death, and the ministry of the Word delivered him from supernatural death. Both represented the highest expressions of God's love and mercy for mankind. It was the duty of the Christian to carry on Christ's ministry in the lights of grace and of nature. His faith must be seen to be justified in his works. Paracelsus put it most succinctly when he wrote: "When Christ spoke and taught, his words were always accompanied by deeds. It should be the same in medicine."[7] As we have seen, Croll encompassed the religious dimension of Paracelsianism within a more economical theology of the Word. Although this deemphasized the personality of Christ, it in no way diminished the religious mission of the physician. For Croll, the two ministries became one, not a ministry devoted to the spreading of the Word by preaching, but a ministry devoted to the revelation of the Word as power.

[6]Ibid., p. 153.

[7]*"Das zweite Buch der Grossen Wundarznei,"* in *Theophrast von Hohenheim, genannt Paracelsus, Sämtliche Werke. I. Abteilung: Medizinische, naturwissenschaftliche und philosophische Schriften,* ed. Karl Sudhoff, 14 vols. (Munich: R. Oldenbourg, 1922–33), 10:281 (hereinafter to be cited as *Paracelsus, Sämtliche Werke*).

In Croll's case the key to this experience and witness of the Word in nature is the Book of Nature itself. This metaphor of the book, however, is a deceptively simple one. For us who have been brought up in a culture dominated by the printed page, it connotes most immediately the idea of nature as a book which can be read and which, moreover, is readily accessible to all. But to the Paracelsian enthusiast, with his innate distrust for the written texts of men, the Book of Nature was not a book like any other: it was *the* book, the companion to the only other book which mattered, the Book of Scriptures. In addition, neither of these two books was an open text of words to be read and cursorily analyzed; they were books of signs whose hidden meanings were to be interpreted. Croll's exegesis, it will be remembered, was Cabalistic—the meaning of the Scriptures was not to be found primarily in the surface discourse of the text but had to be divined in the Hebrew characters and signs of the supposedly pristine text. It was at this level that the power of the Word was manifest. The Book of Nature operated in the same way. Its text was not a discourse of words obeying laws of grammar and syntax but a codex of secret signs whose interpretation must be guided by the light from within and without. In discussing nature as a repository of signs, Croll paraphrases pseudo-Dionysius in the following way:

> Dionysius saith that we cannot know God from his own Nature, but from that most orderly disposing of all the Creatures proceeding from himselfe, which (creatures) hold forth as it were images and similitudes of his Divine Presidents or Examples.[8]

Nature, too, reveals the power of the Divinity. But what is to be noted here is that God is apprehended through the *signs in* nature, and not simply by the *design of* nature. God is immanent in His creation and is not transcendent. The appeal of the signs is to the imagination and not to reason. The Paracelsian Book of Nature is in no sense a natural history. The author of the book is God, who speaks to man in mysterious ways. Man's language consequently must strain to capture the density of meaning conveyed by the signs. It is this very fact which makes Paracelsian texts themselves difficult to interpret. Their fantastical vocabulary is not designed to define unique, singular characteristics of phenomena; rather, it is constructed to reveal as many depths of meaning as possible—their words are intended to reverberate in the imagination with meanings.[9] Frequently men's words fail to communi-

[8]*Admonitory Preface*, p. 50.

[9]On this point, see Owsei Temkin, "The Elusiveness of Paracelsus," *Bull. Hist. Med.* 26 (1952): 201–17, esp. p. 216. This perceptive paper gave me many insights into the problems discussed in this chapter.

cate the truths of imaginative experience, and the Paracelsian is left literally wordless; he can then only beckon to experience. As Croll puts it:

> Therefore the true and more profound Phisitians, who have been Divinely inspired . . . shall use these Medicines rightly prepared by their own labour . . . and not trusting . . . as many doe, to the Sophistical and fraudulent preparations of others, they will know by experience far greater efficacies and operations by the cures they doe, then I can or ought to set down and assign. . . .[10]

The Word in nature which the Paracelsians sought to articulate was, in the last analysis, ineffable.

The Semiology of Experience

In this light it is little wonder that modern commentators have confessed to finding little connection between the "theoretical" *Admonitory Preface* of Croll's *Basilica Chymica* and the "practical" section, where he describes his chemiatric medicaments. The *Admonitory Preface* is not linked to the prescriptive part by rational dialogue; rather it contains an account of cosmogony, delineating man's place and function in the scheme of nature and giving in broad outline the harmonies and sympathies operative within nature in terms of its vital anatomy. In short, it tells one how and where to look for the signs of disease and the cure, but it does not in any detail tell how to interpret these signs—it does not even specify very precisely their character. Further guidance to the links between Croll's *Admonitory Preface* and the description of his preparations is however provided by the third part of the *Basilica Chymica*, the treatise on the doctrine of signatures.[11] Ostensibly a treatise on the signatures of plants, it nevertheless ranges more widely to include the signs of all created species, including man himself, as well as the signatures of diseases and medicaments. It offers an entrée into the world of signs and similitudes which governed

[10] *Admonitory Preface*, pp. 7–8.

[11] The treatise on signatures in the 1609 edition of the *Basilica Chymica* has a separate title page, which reads *Osualdi Crollii Tractatus de Signaturis Internis Rerum, seu de Vera et Viva Anatomia Majoris et minoris mundi.* All subsequent references will be to this edition of the treatise, which was not included in Pinnell's English translation of 1657, and the translations offered are my own.

Croll's thinking and offers us further insights into how to read the Book of Nature.

The treatise on the doctrine of signatures opens with an attack on the botanical nomenclators of the sixteenth and seventeenth centuries:

> Oh that the Botanists of our time, who being ignorant of the internal Form of plants, know only their matter, substance, and body, would devote as much care to the discernment of the Signatures of Plants as they do to their manifold and frequently frivolous disputes about the accurate naming of them, it would render a much richer and more beneficial service to medicine.[12]

Such activity, continues Croll, only results in a multitude of taxonomists concerned exclusively with the external features of plants and with their habitats. He, on the other hand, intends to cultivate the "footprint," the "shadow," and the "image" which the Creator has cast on His creatures and to seek out that internal power and secret virtue which was enclosed there by means of the signatures and the mutual analogy and sympathy of the members of the body and plants. Then he will be able to extract by fire and the scalpel that hidden virtue which those others pass over in somnolent and culpable silence.[13]

The opposition to the taxonomic enterprise here expressed is absolute. It reverses what should be the true relationship between man and creation. Man ought not to address nature with names of his own fashioning; rather, he should let nature speak to him through its own language of signs and so reveal its divine purpose and activity.[14] Humanly constructed names do not convey the interior virtues of things. Only Adam in his state of innocence in the Garden had so perfect a knowledge of natural things that he was able to endow every object with a name which expressed its internal nature.[15] He was able to do so, however, not because he had dominion over nature but because he could read and give perfect expression to the signs. Postlapsarian man lacks such power of denomination, but he must continue to allow nature to transmit its messages to him. This language of creation is most remarkable and providential, muses Croll, for if we

[12]*Tractatus de Signaturis*, p. 1.

[13]Ibid., pp. 1–2.

[14]"Tacentibus nobis loquitur veluti notis quibusdam Natura, ac ingenium cujusque & mores revelat . . ." (ibid., p. 3).

[15]"Primus noster Protoplastes Adam, in statu Innocentiae ex Arte praedestinata, id est signata, rerum Naturalium absolutam habuit cognitionem . . ." (ibid., p. 15).

were to imagine that plants could speak, they would have to be multilingual in order to convey their messages to all of mankind; but with their universal language of similitude, they convey their meaning concisely and clearly to all. All plants, flowers, trees, and the other products of the earth are texts or magic signs conveyed to us by the infinite mercy of God to signify where true medicine lies hid. The signs themselves are not the medicine, but signatures pointing to it.[16] All creatures thus become teachers of medicine—the divine study of their signatures reveals more and more for the art of medicine, a fact which Croll admits some botanists do acknowledge, preferring, however, to leave the teaching and demonstration of it to others.[17] Anyone who is ignorant of the philosophical and medical alphabet of nature cannot be a genuine physician, for these characters and natural signatures are not written in ink but have been inscribed by the finger of God on his creatures.[18] It is a study which does not require great deductive reasoning; only experience, the mother of truth, merits confidence.[19] Principally it requires that man stand in silence before the voice of nature, which is in turn the voice of God.

Another aspect of contemporary botany and materia medica which Croll deprecates is the fascination with foreign species and exotica at the expense of domestic species. Although the natural language of signs is universal, there is no uniform distribution of created species. Instead God, as a sign of His beneficence, has provided each region of the earth with a bountiful supply of medicinal sources with virtues appropriate to it:

> Just as the Earth provides sustenance and clothing for every region, if not to the extent of superfluous luxury, at least sufficient to meet necessity, so also does the mother of things, Nature herself, provide abundantly for all and distribute a necessary sufficiency of medicaments. Each country contains within itself the matrices of its own Element, and these provide what is necessary for it. For every land and region, Nature brings forth plants tempered according to its people, its climate, its heaven, and its age.[20]

Thus in Moravia, where the stone, gout, and contraction of the limbs

[16]Ibid., p. 3.

[17]Ibid., pp. 14–15.

[18]"Et qui hujus Fundamenti intellectu caret, hujusq[ue] Philosophici & medici Alphabeti cognitione destituitur, non potest esse probatus medicus. Naturae enim characterismi, & hae Signaturae Naturales, quae ex Creatione non atramento, sed ipso DEI Digito in omnibus sunt exarata . . ." (ibid., p. 15).

[19]"Magna Rationum deductione non egemus, si Experientia Veritatis mater Fidem meretur" (ibid., p. 2).

[20]Ibid., p. 5.

are common afflictions on account of the local sedimentary wines, nature has provided a number of suitable remedies which are peculiar to the region.[21] Such ideas reflected the basic Paracelsian doctrine of the specificity of pathological and therapeutic species, as well as the concept of the *anatomia essata*. There also echoed from a distance the ancient Hippocratic notion of the influence of climate and topography on temperament and disease.

As a further example of God's bounty in the provision of remedies, Croll cites His providence in furnishing a mineral alternative to the unicorn's horn, an exceedingly rare and hence costly medicament. In connection with the unicorn Croll recounts a curious episode from his own experience. While practicing medicine at Brno with Dr. Johann Berger, Croll witnessed the digging up, from under a very high rock, of the skeletons of two extraordinarily large and unknown species of animal, together with the bones of two immature specimens. Croll believed these to be relics of the Flood, and visiting the site some months after their discovery, he took away some bones and teeth of the specimens, which he found to have similar medicinal effects to those attributed to the horn of the unicorn. The moral of this strange discovery, although unstated, is quite clear: God has provided even the Moravians with the equivalent of the most rare of medicinal species; all that is required is a most diligent search and conscientious reading of the signs on the part of the natives.[22] Croll cites with approval Paracelsus's mirth at the exaggerated curiosity of many physicians towards foreign plants and their names, while the peasant has a veritable pharmacopoeia at his doorstep.[23] This critique of the obsession with exotica passes over into a hostile social commentary on the fastidiousness and greed of contemporary physicians and apothecaries:

> The local [medicaments,] which God has given us in sufficient supply for the afflictions of the body, have begun to go to waste, because they require a diligent and devoted confidence in their preparation, from which the silk-clad physicians shrink on account of their aversion to the dirt of coals, smoke, and clay. And also because the greatest proportion of general apothecaries, eager for the incentives of fame and greed, are preoccupied in their shops and are more intent on emptying the purse of the sick person than on restoring his body.[24]

The world of Croll takes its meaning from the signatures which are imprinted on it. These signs were indications of the appropriateness

[21] Ibid.
[22] Ibid., pp. 4–5.
[23] Ibid., p. 5.
[24] Ibid., pp. 6–7.

of things—of their *convenientia*—revealing the basic sympathies and antipathies which held the world together. Ideally the signatures were the means of tracing out the elements of the golden chain which linked all things in the universe, from the lowest element to the Divinity itself. All things had their proper interconnection. Everything stemmed from the creative fiat of the Father, and every specific power and object was a particular manifestation of this one Word in its own degree of corporification. In theory, by reading the signs correctly, man could ascend from a knowledge of natural objects to a knowledge of the Divine. But matters were not so simple in practice. The chain of *convenientia* wound like a labyrinth through all of nature: the signs reflected their sympathies and antipathies back on one another with myriad complexity. The task of interpreting them would have proved impossible if not for the fact that all found their way back to man, who as the microcosm was the focus of all similitude in the great world.[25]

All the sympathies and antipathies of the great world find their true correspondence with man, and all the signatures lead back to him as the microcosm. Physiognomy and the face of the sky; the lines of man's hands and features (chiromancy) and the veins of minerals in the earth; the pulse and the regular movement of the heavenly bodies;[26] the seven members of his body and the seven metals and the seven planets; man's bones and the species of wood—these are the major correspondences.[27] Man's afflictions too are signified in the great world—fevers by earthquakes; colic by thunder; apoplexy by eclipses and thunderbolts; epilepsy by storms.[28] Croll even offers an elaborate prognosis of an attack of the grand mal with reference to the passage of the storm.[29] It is the fundamental macrocosm-microcosm correspondence which binds the cosmos together. Man is the center which holds.

Although Croll refers to the signatures as the alphabet of nature, the metaphor of the book is an inadequate one for the experience he wishes to convey. The signs in nature are more than mere ciphers of a printed text. They appeal to more than just the visual sense of man; they engage all the senses, and the emotions, and the actions of

[25]In the following discussion my debt to Michel Foucault, *The Order of Things: an Archeology of the Human Sciences* (New York: Random House, Pantheon Books, 1970) will be obvious; particularly to chapter 2, "The Prose of the World," where Croll is included in the discussion. However, I would argue that Foucault interprets the metaphor of the Book of Nature too literally; see below.

[26]*Tractatus de Signaturis,* p. 52.

[27]Ibid., p. 54.

[28]Ibid., p. 52.

[29]Ibid., p. 53.

men—in short, they embrace all experience. Hence the feel of the callused pedicles of the roots of ivy is a sign of its virtue for scrofula;[30] those who laugh a lot are signed by the cry of the macaw;[31] the roots of the orchid are signed as an aphrodisiac because of their smell.[32] The visible signs, too, appeal to all dimensions of the visual sense. Rhubarb, because of its color, expels the yellow bile;[33] crabs, on account of their shape, provide remedies for skin cancers;[34] and the lily-of-the-valley, which droops its head, has a virtue in cases of apoplexy.[35] The poor victim of hydrophobia is signed by the fear of water, but therein lies the cure: push him in when he is not looking![36] The therapeutic action is signed by the pathological fear. But the strangest sign of all is the sign of silence. In the preface to the *Tractatus de Signaturis,* Croll notes that Moses did not mention the precious stones and metals in his account of creation, although he recounted everything that was visible on the surface of the earth. His silence was a sign that God had hidden the most powerful arcana of nature out of sight, below the surface of the earth. Their location there, in the center of the great world, was also a sign that their great virtues were designated for the preservation of the vital balsam of the heart of man, the center of the microcosm.[37] Plants on the surface of the earth had in contrast much weaker virtues and were intended for the preservation of the peripheral members and features of man. Here we meet the most frustrating aspect of the Paracelsian therapy. Its most notable innovation was the advocacy and chemical preparation of metallic and mineral medicaments, whose great virtue was extolled. But because these were the most powerful of arcana, they were surrounded by the greatest aura of reticence. Croll restricts himself to revealing the correspondence between the metals and the planets together with their conventional alchemical symbols, and even this he regards as the revelation of a great secret.[38]

[30]Ibid., p. 38.
[31]Ibid., p. 62.
[32]Ibid., pp. 23–24.
[33]Ibid., p. 45.
[34]Ibid., p. 43.
[35]Ibid., p. 27.
[36]Ibid., p. 45.
[37]Ibid., p. 9.
[38]Ibid., pp. 77–78.

Signs in Practice

Such, then, is the nature of the intuitions which governed Crollian diagnosis, prognosis, and therapy. There remains, however, the difficult task of elucidating how these insights found expression and application in the preparative section of the *Basilica Chymica.*[39] At first reading, there appears to be a disconcerting lack of confirmation of Paracelsian principles—set out in such detail in the *Admonitory Preface*—in the actual prescriptions of Croll's pharmacy. The expectations of a detailed and specific therapy, aroused by the tenets of pathological specificity based upon the chemical anatomy of the great and little worlds, do not appear to be wholly fulfilled in the remedies which Croll describes. In the first place, many of the innovative chemical preparations introduced by Croll, particularly those resulting in new compounds of antimony and mercury, are included in the first category of remedies which describe emunctories. This group of prescriptions is classified under the traditional headings of emetics, cathartics, diuretics, and diaphoretics, thus seemingly preserving the Galenic emphasis on drugs which expelled excess and harmful humors.[40] Croll argues, however, that his evacuative remedies operate in a specific way, not by the qualities of the four humors, but by the "occult property" and "specific form" of the medicament which drives out the humor and the excretion peculiarly associated with it.[41] The use of such drastic ingredients as mercury, antimony, and vitriol in the preparation of cathartic drugs is justified in terms of their ability to remove the "fixed roots" of disease in the body. Just as these latter are tenaciously seated in the depths of the microcosm, so potent remedies from the bowels of the macrocosm are required in order to remove them. Croll lauds the new spagyric pharmacy for its ability to refine chemically these potentially dangerous mineral species into forms which could be administered safely.[42]

[39]In the 1609 edition of the *Basilica Chymica, continens philosophicam propria laborum experientia confirmatam descriptionem & usum remediorum chymicorum selectissimorum è lumine gratiae et naturae desumptorum. In fine libri additus est eiusdem Autoris Tractatus novus de Signaturis Rerum Internis* (Frankfort) the preparative section is continuously paginated with the *Admonitory Preface.* It runs between pages 111 and 283. All subsequent references will be to this edition.

[40]These remedies are described in the *Basilica Chymica* (1609), pp. 114–84.

[41]Ibid., pp. 121–22.

[42]Ibid., p. 122.

In the second major group of remedies, in which Croll describes preparations for the seven principal parts of the body, he does appear to go further in the direction of pathological and therapeutic specificity, at least as regards classification of his medicaments.[43] However, this is in part negated by the fact that this class of remedies, like the emunctories, contains many polypharmaceuticals, that is, mixed drugs containing numerous ingredients. Notable in this regard is the first such remedy, prescribed for the preservation of the seven principal members of the body in cases in which the cause and nature of the disease is unknown. It is a veritable arsenal of spagyric pharmaceuticals mixed together and administered as an electuary.[44] Its composition and use would appear to mark a self-acknowledged recognition of the limits of Paracelsian theory in practice.

Finally, there is the question of the range of therapeutic species described by Croll. In the light of the Paracelsian emphasis on the diversity of pathological species signed by the diversity of things that grow in the great world, it is surprising to find that Croll's preparations are based on a comparatively small number of chemical species. Robert Multhauf, who has studied in some detail the development of chemical pharmacy from the fourteenth century on, notes that most of the mineral species used as starting materials by Croll for his preparations—notably mercury, antimony, the vitriols, and gold—were the favored ingredients of chemical pharmacists long before Paracelsus. He depicts Croll's principal innovation, not in terms of the variety of mineral species employed, but rather in terms of the skillful and inventive exploitation of a few traditional resources to provide *reagents* for the preparation of new medicaments.[45] In this light Croll's "chemistry" emerges as an ingenious elaboration of the chemical properties of a few substances, but scarcely as the production of an exhaustive chemical anatomy of the great world.

Are we then justified in completely divorcing Crollian practice from its supposedly underlying Paracelsian ideology? Certainly this is permissible if our sole aim is to delineate Croll's positive contributions

[43]These remedies are described in ibid., pp. 184–226.

[44]This remedy is entitled "Medicamentum specificum omnium septem membrorum principalium." Ibid., pp. 184–86. Among other ingredients the remedy contains spirit of vitriol, oil of amber, salt of pearls, milk of sulphur, diaphoretic antimony, and tincture of the flowers of iron.

[45]Multhauf's analysis of the preparations of Croll is contained in his article "Medical Chemistry and 'The Paracelsians,'" *Bull. Hist. Med.* 24 (1954): 101–26; see esp. pp. 110–23 and 125–26. This analysis is based upon a late (1643) edition of the *Basilica Chymica* and includes many preparations not in the original.

to the development of chemical knowledge and technique. But it should be noted that it is a strategy that succeeds only by isolating that small number of Croll's prescriptions which yield to analysis in modern chemical terms. Moreover, it is one which leads to some uncomfortable conclusions. In terms of such a positivistic analysis, the contributions of the Paracelsians to chemical pharmacy appear to lie in their emphasis on chemically prepared medicaments from mineral species rather than in any really innovative reforms in this regard. What contemporaries regarded as the most radical feature of their practical program is almost explained away.[46] But if we set aside the reagent bottles of the modern chemical laboratory and attempt to understand Croll's prescriptions in terms of his own theory—the theory of signatures—then perhaps we may bridge in part the apparent gap between his theory and practice.

What, for instance, has chemistry to reveal about the recipe for the *Zenexton of Paracelsus?* Here is described in great detail the manufacture of amulets composed of a paste of eighteen powdered toads dried in the sun, together with white and red arsenic, pearls, corals, fragments of precious stones, and herbs. The amulets are stamped on opposite sides with impressions of a scorpion and a serpent, this procedure to be carried out when both the sun and the moon are in the constellation of Scorpio. The resultant objects are wrapped in red muslin and suspended around the neck so as to lie near the region of the heart. This remedy provides protection from the plague and from all harmful astral influences, drawing out the poison from within and devouring that from without. Clearly here we are back in the realm of talismanic magic, astral influences, the Paracelsian doctrine of the plague, and the theory of signatures.[47]

Likewise we are in familiar Paracelsian territory in the preamble to Croll's recipe for potable gold, a "cordial" for the preservation of the heart. Here the correspondences and similitudes of the sun, the center of the macrocosm, its corresponding metal gold, and the heart of man, the center of the microcosm, are fully elaborated. The recipe itself describes the solution of gold in aqua regia, followed by precipitation of the powdered gold fulminate by means of the addition of oil of tartar. An attempt to render this powder potable is made by digesting it in a mixture of spirit of salt and spirit of urine, subliming the resulting salt and suspending it in spirit of urine. For Croll, this was not primarily a triumph of mineral acid technology over refractory gold (the results are ambiguous in terms of modern chemistry), but an effort

46This dilemma is hinted at in Multhauf's concluding paragraph, ibid.,p.126.

47The preparation of the Zenexton Paracelsi is described in the *Basilica Chymica* (1609), pp. 237–40.

to free gold from its terrestrial solidity and to extract its vivifying principle. Thus the striving for a volatile and reactive species.[48]

It is through such signs that we must try to grasp the motivation behind the praxis. Tracking these down is by no means a simple exercise: they frequently elude us, for they may reside in the ingredients, in the technique, and in the characteristics of the product; and many are probably forever lost in some long forgotten analogies. In the case of vitriolated tartar, one of the first chemically prepared medicaments described by Croll, the sign is in the starting material, the process, the nature of the product, and the therapeutic end. A seemingly straightforward preparation of crystalline potassium sulphate, made by dissolving the salt derived from tartar (potassium carbonate) in sulphuric acid from vitriol, it is recommended for calculary diseases and for removing obstructions of the viscera and veins, which according to Paracelsian pathology arise from deposited tartar. In this recipe the pathological species is resolved, purified, and rendered soluble by the processes involved and is confirmed by the solubility of the product. Here the sign runs right through the prescription.[49] As for the fetid spirit distilled from tartar, the sign is in the smell of the product. Although Croll suggests that this can be mitigated, he stresses that it should not be removed entirely, as the pungent medicament in losing its sign will lose its efficacy. It is recommended for expelling putrid sweat from the bodies of victims of the plague.[50] And finally there is the sweet-tasting sugar of lead, which is prescribed for external use against the ulcers and skin eruptions of "saline" diseases. The sign is in the taste. Just as the sugar of the vegetable kingdom mitigates acrid and saline flavors, so this more powerful mineral sugar tempers the excrescences arising from mineral salts in the microcosm.[51]

Of course Paracelsian therapy was bounded by the limits of the chemical knowledge and expertise of its times. In this regard Paracelsian intuition far outran existing knowledge and technical capacities. But there remains another more fundamental reason for our inability to comprehend fully their practice and its connection with underlying theory. This is the conscious awareness of the individual Paracelsian like

[48]The preamble and recipe are in ibid., pp. 201–18.

[49]See the recipe for Tartarus Vitriolatus in ibid., pp. 115–18.

[50]"Foetor ille spiritus Tartari in totum non est auferendus, nec formidandus, cum habeat signaturam Foetorum, quando videlicet in Peste . . . magno juvamine profluentes foetidos sudores è corpore copiose expellit" (ibid., p. 159).

[51]"Sicuti etiam in Vegetabilibus Saccharum vulgare acrimoniā & amaritudinem rerum temperat ac cρrrigat: ita Mineraliū & Mercurij & Arsenici tam in magno quam parvo mundo hoc Saturni Saccharum mitigat ac emendat" (ibid., p. 274).

Croll of the inadequacy of men's words to express his imaginative grasp of reality. He invites us to share his experience, but he cannot fully articulate it for us. It is locked up in the recesses of his imagination and sealed with the sign of ineffability. Here the historian, who necessarily deals in words, is confronted with the ultimate paradox of Paracelsianism. As if to remind us of it, Croll ends his *Basilica Chymica* with a prayer to Sacred Silence.[52]

[52]This occurs at the end of the *Tractatus de Signaturis*, pp. 75–76.

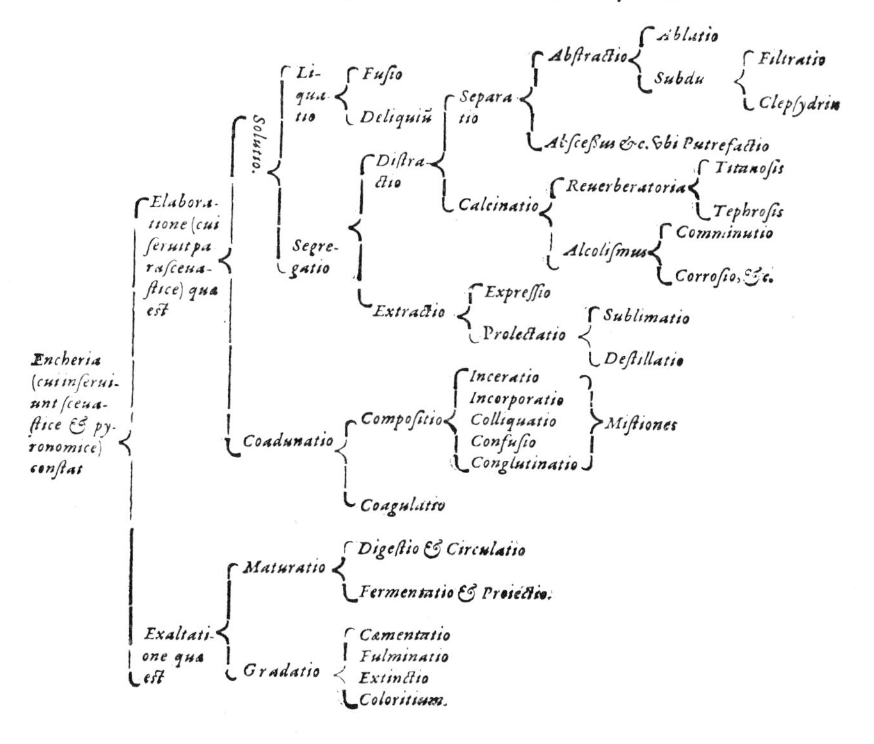

TABVLA PRIMI LIBRI Alchemiæ.

Alchemia habet partes duas Encherian, & Chymian.

- Encheria (cui inseruiunt sceuastice & pyronomice) constat
  - Elaboratione (cui seruit parasceuastice) qua est
    - Solutio.
      - Liquatio
        - Fusio
        - Deliquiũ
      - Segregatio
        - Distractio
          - Separatio
            - Abstractio
              - Ablatio
              - Subdu
                - Filtratio
                - Clepsydria
            - Abscessus &c. Ubi Putrefactio
          - Calcinatio
            - Reuerberatoria
              - Titanosis
              - Tephrosis
            - Alcolismus
              - Comminutio
              - Corrosio, &c.
        - Extractio
          - Expressio
          - Prolectatio
            - Sublimatio
            - Destillatio
    - Coadunatio
      - Compositio
        - Inceratio, Incorporatio, Colliquatio, Confusio, Conglutinatio — Mistiones
      - Coagulatio
  - Exaltatione qua est
    - Maturatio
      - Digestio & Circulatio
      - Fermentatio & Proiectio.
    - Gradatio
      - Cæmentatio
      - Fulminatio
      - Extinctio
      - Coloritium.

TABV-

Plate 2. The Words of Chemistry. Plates 2 and 3 (p. 74) are dichotomized summaries of chemistry from Andreas Libavius's *Alchemia* (1597). These wholly verbal tables of the chemical art should be contrasted with the pictorial symbolism of Croll's title page (plate 1). Plate 2 sets out methodically the operations of chemistry (the *encheria*). (Courtesy of the University of Wisconsin Library.)

TABVLA LIBRI SECVNDI ALCHEMIÆ.

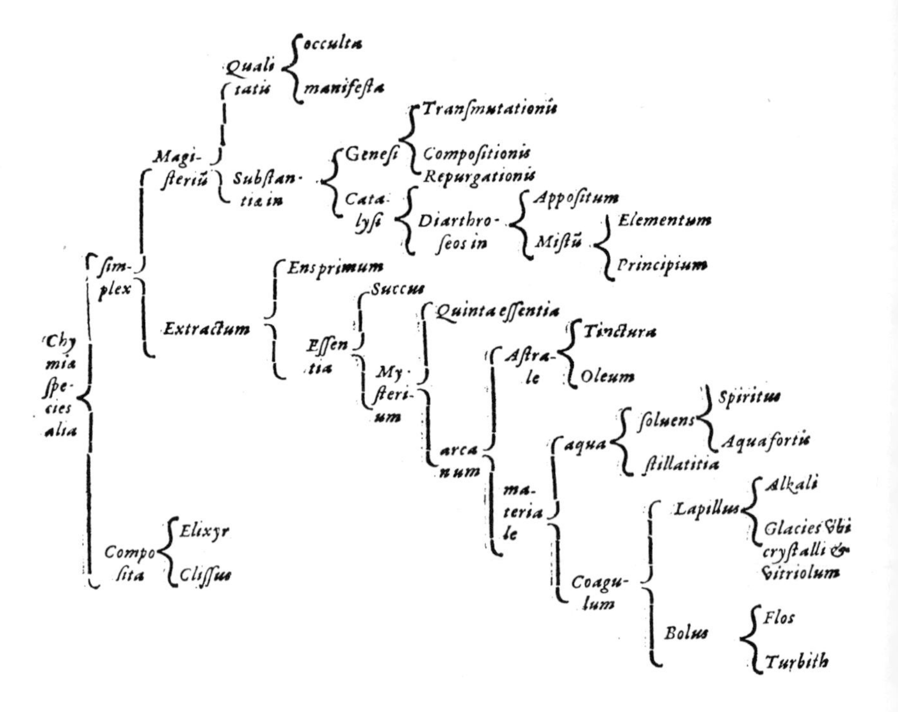

PHA

Plate 3. The verbal classification of chemical species (Libavius, *Alchemia,* 1597). (Courtesy of the University of Wisconsin Library.)

CHAPTER IV

CHEMISTRY AS COUNTERCULTURE

The strong conservative reaction which Paracelsianism engendered is exhibited in a letter written by the pious and conscientious schoolmaster Andreas Libavius to a young physician friend, warning him of the evils of chemistry. The letter, published in 1595, was the first in what was to become a three-volume series of open epistles on chemical topics, written to philosophers and physicians throughout Germany.[1] The tone of the letter is one of solicitous paternal pleading: Libavius seeks to warn his young friend, who appears to have been his former pupil,[2] of the potential dangers to his health, social position, and scholarly reputation which he is courting by having dealings with chemical philosophy. He attempts to expose for his protégé the impieties and follies of the chemists' claims, their antisocial behavior, and their overbearing arrogance. The letter does not contain a detailed intellectual attack on Paracelsian doctrine; it is more urgently personal in its concerns. The disconcerted professor, a pillar of the academic establishment, is seeking to rescue a promising pupil from the seductions of the counterculture of his times.

[1]*Rerum Chymicarum Epistolica Forma ad Philosophos et Medicos quosdam in Germania excellentes descriptarum Liber Primus, in quo tum rerum quarundam naturalium continentur explicationes ingeniosae; tum Chymiae disciplina pyronomica, sceuastica & vocabularia cum quibusdam inter arcana habitis declarantur fideliter. Autore Andrea Libavio Med. D. Poeta & Physico Rotemburgo tuberano* (Frankfort, 1595), hereinafter to be cited as *Rerum Chymicarum Epistolica.* The second volume also appeared in 1595 and is sometimes found bound with the first volume. The third volume was published in 1599.

[2]The name of the young friend is given as Cosman Crenandrum, described as a physician at Stade. However, in the epistle which follows, and which from the context is obviously addressed to the same person, his name is given as Cosman Pegandrum. This would appear to be the result of a misprint, as there is no reason to believe that Cosman was a fictitious correspondent. Most of the letters were directed to well-known living philosophers and physicians in German-speaking territories. Libavius had an extensive correspondence, much of which remains unpublished. See W. Hubicki, "Libavius, Andreas," in *D.S.B.*, vol. 8, 309–12.

Libavius recalls his pupil's serious prosecution of philosophy and medicine, noting that he was not content to rest on his laurels but sought to understand the more abstruse secrets of nature. Alas, he has fallen in with the chemists!—the chemists, who promise a hidden wisdom not appropriate for everyone; who appear to debate with one another openly using normal language and writing but who in fact conceal their true meaning in letters which the philosophers of Egypt used to call hieroglyphs. He has given himself over totally to this science and plunged into it to such an extent that he has altogether disappeared from view. Just as once his old professor commended him for his philosophical ardor and thirst for medicine, now he seeks to warn him not to accept everything the chemists write without subjecting it to the most serious critical judgment.[3] Libavius reminds him of his religious and social heritage: "We were born to the religion of the true God and to nurture, assist and protect human society." Anyone who takes himself off to the chemical furnaces deserts these obligations. Those who indulge in such activity forsake the reading of the Scriptures, hold the law in contempt, and neglect their domestic duties, while they spend their lives with flames and burning coals, struggling with smoke.[4]

Libavius goes on to attack the chemists' pretensions and impieties in claiming to perfect nature. Arguing from an Aristotelian standpoint, he denies that nature generates imperfect things; the works of nature are perfect in themselves, and any blemish stems from factors outside of nature. The tendency towards perfection lies deep in nature itself, and unless the chemists have the power of balancing the qualities, they cannot claim to produce anything more perfect than nature. But what in fact do the chemists do in their furnaces? They destroy the balance of qualities in the *mixis* and reduce everything to formless matter; and surely no one is so impious and mindless as to suggest that the chemist can endow matter with forms at will and so generate new species. If they obtain anything, it is purely by luck.[5] By contrast, nature is so abundant in its riches that, in spite of the innumerable

[3]*Rerum Chymicarum Epistolica,* 1: 1–4.

[4]"Nati & veri Dei religioni sumus, & humanae societati colendae, iuvandae, tuendaeque. Stationem hanc deserit, parumque sortis suae memor est, quisquis in solitudinem fornacum chymicarum ingressus inter Vulcani latet tripodas, & quasi nec divinorum officiorum nec humanorum rerum ad se pertineret cura, posthabita sacrarum literarum lectione, iuris civilis inspectione repudiata, & legum domesticarum sanctione neglecta, cum flammis prunisque conversatur, colluctatur fuliginibus & nidorum acumine sensus perdit" (ibid., pp. 2–3). This and all subsequent translations from Libavius's works are my own.

[5]Ibid., pp. 4–7.

generations of men who have studied her, there is still more to be known. There is more than enough to occupy a lifetime in this study. There are the writings of the ancient philosophers, so full of abstruse knowledge and sagacious discovery.[6] Libavius reminds his pupil that he has witnessed in the universities of Germany, Italy, and France the great task involved in explicating the meaning and sentences of the philosophers of antiquity. It is to this great labor of humanistic scholarship that he wishes to recall his former student.[7] He does not wish to see him waste his intellectual abilities by involving himself in chemistry to the detriment of his health—all those poisonous vapors and night-watches over the furnaces—his duty to God, his country, friends, and relations.[8]

Libavius goes on to remind his pupil of the legal dangers he is courting by allowing himself to be classed as a chemist. He will be regarded as a magician. Throughout history legal opinion has been against chemists. Their books have been burned; in almost every nation chemists have perished—witness that famous physician Arnald [of Villanova], who after a cruel imprisonment just escaped a more cruel death. Moreover, chemists frequently suffer at their own hands: while they gaze with longing at mercury undergoing transmutation, they more quickly turn themselves into beggars and outcasts of society.[9]

The chemists also claim to aid medicine, but look at their works: they administer such poisons as antimony and mercury even internally, and they dose their patients in great quantity with a water extracted from tartar and nitre, which they mix with aqua regia. Erastus has not hesitated to call their medicines diabolic.[10] In reply they have asserted that their medicaments have been drawn from salts and juices which have been added to food since antiquity. But it is easy to see the stupidity of this, for the waters which they extract from safe

[6]Ibid., p. 7.

[7]"Multas vidisti academias Germanas Italasque & Gallicanas. In qua non laborem illum laboriosissimum, quo explicantur saltem veterum sensa & phrases, deprehendisti? . . Quo vultu ergo ausit quis posthabitis illis omnibus chymian irrepere?" (ibid.).

[8]Ibid., p. 8.

[9]Ibid., pp. 9–10. The allusion to Arnald of Villanova most probably refers to the condemnation of his theological views at Paris in 1299, when he was spared on the intervention of the French crown. See Michael McVaugh, "Arnald of Villanova," *D.S.B.*, vol. 1, p. 289.

[10]Ibid., pp. 11–12. Thomas Erastus (1523–83), the Swiss physician, philosopher, and theologian, was one of Paracelsus's most hostile critics. His *Disputationes de Medicina Nova Paracelsi* were published in four parts in 1572 at Basel. See Walter Pagel, *Paracelsus: An Introduction to Philosophical Medicine in the Era of the Renaissance* (Basel and New York: S. Karger, 1958), pp. 311–33.

and good things acquire a bitterness and poisonous acrimony from the fierce fires they use in their preparation. They have so little confidence in their own liquors that they proffer them in the darkest of phials and only dare to administer them in the smallest doses and with great dilution.[11]

As for the chemists' claim that they are philosophers, what vanity! The more distinguished a philosopher is, the more he keeps his distance from the chemists. The ancient philosophers, such as Plato and Aristotle, who have dealt with everything concerning philosophy, have said as much about chemistry as they have said about charcoal.[12] Later philosophers to whom the art was known have condemned it. In none of the universities of Europe is it taught or thought worthy of forming a part of liberal studies. No one would teach it from a public platform unless he wished to be proclaimed a teacher of fraud and deceit and a trumpeter of barbarism.[13] Chemistry is the occupation, not of philosophers, but of reprobates. Libavius again asks his pupil rhetorically whether he wishes to give up his place amongst philosophers to go along with such people.[14]

Finally comes an attack on the chemists' allegiance to Hermes and on their secrecy. If Hermes had been a good philosopher, one who wished to be consulted by posterity, he would not have cloaked his obscure teaching in a fog and left his reader with unexplained perplexities thus giving rise to a whole colony of imposters to interpret him. Moreover, he arrogantly and impiously took the name Trismegistus, which was previously attributed to the god Mercury. If you look to his followers, you will find more of the same obscurity and Egyptian or Cimmerian darkness. Egypt, from where Hermes came, was always suspected of being the home of magic and idol worship. The children of Egypt and Hermes display the same characteristics.[15] Thus chemists cannot abide critics; they guard their art like the Eleusinian mysteries. If you wish to be one of their disciples, you are sworn to silence, as if joining a band of rogues and assassins.[16] Libavius ends with the

[11] *Rerum Chymicarum Epistolica,* 1: 12–13.

[12] Ibid., pp. 14–15.

[13] "Tot sunt academiae in universa Europa. Nullibi valet eius artis professio, nempe indigna iudicata est liberalium & ingenuorū studiorum conditione. Et quis eam in publico sapientiae suggesto pulpitisq́; doceret, ni vellet fraudum & imposturarū magister, imò barbariei buccinator proclamari?" (ibid., p. 15).

[14] Ibid., p. 16.

[15] Ibid., pp. 16–17.

[16] "Si discipulum profitearis; oportet te silentium iurare, non aliter ac hi qui in improbissimorum nebulonum sicariorumq́ue admittuntur gregem" (ibid., p. 17).

caution: "Beware of deceiving chemistry, and take my advice, keep away from it!"[17]

The author of this diatribe against the chemists, Andreas Libavius, was one of the most prolific controversialists of Germany at the turn of the seventeenth century. His polemical prowess and energies engaged some of the major intellectual controversies of his time, including Ramism, Hermeticism, Paracelsianism, Jesuit theology, and Rosicrucianism. His literary career spanned the years 1591 to 1616, during which time he was the author of some forty-six works, many of considerable bulk and nearly all published at Frankfort on the Main on poor quality paper which has not withstood well the ravages of time. As a dedicated polemicist, Libavius was not content with a short tractate where a folio of some five hundred pages would equally serve his purposes. Intimidation by erudition might best describe his tactic. Irrespective of its effectiveness in his own time, it has certainly succeeded as far as posterity has been concerned: the wordy Libavius has had remarkably few words written about him; the academic enterprise to which he devoted his whole life has paid him back in small change indeed.[18]

This is hardly surprising. The sheer size of his more important treatises would be enough to frighten off the most dedicated pedant, and the range of subject matter—logic, theology, medicine, chemistry, pansophy, the magical arts—would seem to belie any inherent unity to his intellectual endeavor, at least to the modern mind. Furthermore, a cursory perusal of the contents of his writings would seem to indicate an intellect which adopted conflicting opinions on the intellectual

[17]"Tu tibi de impostoria chymia cave, & me monitore ab illa abstine" (ibid., p. 18).

[18]The best informed source on Libavius is Hubicki, "Libavius, Andreas," in *D.S.B.*, vol. 8, pp. 309–12. Other valuable assessments of his work from a strictly chemical point of view are: Ernst Darmstaedter, "Libavius," in G. Bugge, *Das Buch der Grossen Chemiker,* 2 vols. (Berlin: Verlag Chemie, 1929–30), 1: 107–24; J. R. Partington, *A History of Chemistry,* 4 vols. (London: Macmillan, 1961–70), 2: 240–77; and R. P. Multhauf, "Libavius and Beguin," in *Great Chemists,* ed. E. Farber, (New York, 1961). L. Thorndike, *A History of Magic and Experimental Science,* 8 vols. (New York: Macmillan, 1923 [vols. 1–2]; Columbia University Press, 1934–58 [vols. 3–8]), 6: 238–53, deals with a broader spectrum of Libavius's polemical activity, but not very perceptively. Other more specialized items on Libavius will be cited in the course of the discussion which follows. The most comprehensive bibliography of his works, some of which are exceedingly rare, is contained in the modern German translation of his masterwork, *Die Alchemie des Andreas Libavius: ein Lehrbuch der Chemie aus dem Jahre 1597,* trans. with illustrations and commentary under the supervision of Friedemann Rex, for the Gmelin Institue for Inorganic Chemistry, in association with the Society of German Chemists (Weinheim: Verlag Chemie, 1964), in the commentary under the heading *Das Corpus Libavianum.* Partington also gives extensive bibliographical information.

matters which concerned it. Thus his first published work was an attack on two British Ramist philosophers, William Temple and James Martin;[19] yet he was a Ramist of a type himself, incorporating elements of the reformed logic into his own textbook on the subject.[20] He was a vituperative opponent of Paracelsianism, yet he defended two of France's most noted Paracelsians, Joseph Duchesne (Quercetanus) and Turquet de Mayerne, against the Paris Medical Faculty.[21] He attacked Hermeticism and Rosicrucianism,[22] but he defended alchemy.[23] And his aforementioned letter on the evils of chemistry notwithstanding, he devoted much of his life's work to the exposition of the art of chemistry and provided the subject with its first definitive textbook, his *Alchemia* of 1597.[24] What is one to make of such a man? Perhaps nothing more than that he loved controversy for controversy's sake; that his polemical skills were so tempered that he could employ them on both sides of a question with the consequence that his intellectual efforts were exhausted in a frustrating search for a middle way through the academic disputes of his time. There is probably much truth in this characterization; but as I shall attempt to show in part, there is an inherent consistency in Libavius's intellectual posture, which gives some coherency to his multifarious intellectual endeavors.

But what of the man himself? In contrast to his writings, the known facts about his life are few in number, although they have an appropriate prosaic quality. He was born around the middle of the sixteenth century in Halle, Saxony, the son of a linen weaver. Educated at the University of Jena, he graduated doctor of medicine and doctor of philosophy there sometime prior to 1581, when he took up a position as schoolmaster in Ilmenau. From there he crossed the Thuringian Forest to Coburg in 1586, where he spent two years as rector of the town school. Progressing steadily up the academic ladder, he taught (classical) history and poetry at his own University of Jena from 1588 to 1591. In this last year he moved to Rothenberg on the Tauber,

[19] *Das Corpus Libavianum,* in *Die Alchemie,* item 1.

[20] Ibid., items 6 and 36.

[21] Ibid., item 34.

[22] Ibid., item 45.

[23] Ibid., item 27.

[24] *D.O.M.A. Alchemia. Andreae Libavii Med. D. Poet. Physici Rotemburg. Operâ & dispersis passim optimorum autorum, veterum et recentium exemplis potissimum, tum etiam praeceptis quibusdam operae collecta, adhibitisq; ratione & experientia, quanta potuit esse, methodo accuratâ explicata, & in integrum corpus redacta . . .* (Frankfort, 1597). In the second edition, which is a handsome folio with commentary, *Alchemia* is tranformed into *Alchymia.* It was published at the same house in Frankfort in 1606.

where he was *stadtphysikus,* a sort of sixteenth-century inspector of public health, a position he combined with that of inspector of public schools from 1592. In 1607 he returned to Coburg as director of the newly founded Gymnasium Casimirianum, a post he held until his death in 1616.[25] It was a life severely limited in its geographical horizons and intellectual experiences. Yet with this career of modest provincial accomplishment as a base, Libavius sallied forth in his writings to do battle with those intellectual and cultural forces which threatened the very fabric of the society that sustained him. We have already encountered some of the virtues which Libavius sought to defend—a duty to God, country, home and society, respect for the law, and the maintenance of academic standards. These are the values of a dedicated pedagogue. Libavius's principal mission in life was to keep the world safe for schoolmasters.

The principal focus of what follows will be Libavius's contributions to the definition and systematization of the chemical arts, although I shall attempt to demonstrate the relationship of some of his other intellectual concerns to this endeavor. Andreas Libavius's achievements in chemistry have for a long time evoked the admiration of historians; but his position, influence, and significance for the ongoing development of the subject have, by and large, defied assessment. His principal work on chemistry, the *Alchemia,* published in Frankfort first in 1597 and in expanded form in 1606, is recognized as the most comprehensive synthesis of chemical technique and prescription to have appeared in print up to that time. Its ambitious scope, its mastery of the accumulated literature on the chemically oriented crafts, its rigorous organization, and above all its forceful statement and attempted demonstration of the independence and integrity of chemistry is without comparison in the seventeenth century. But both the motivation behind this accomplishment and the intellectual principles and ideals which informed it have not been subjected to any detailed analysis. Pagel, in a brief review of the recent German translation of the *Alchemia,* has pointed up the Aristotelian character of the organization and presentation and has suggested that the book belongs in spirit to the humanist form of Aristotelianism prevalent in the mid-sixteenth century.[26] But by and large, only particular elements of Libavius's chemistry have attracted the attention of recent scholars. Partington

[25]Biographical information is taken from sources in n. 18, above. Hubicki, in "Libavius, Andreas," *D.S.B.,* vol. 8, p. 309, states that Libavius "enrolled" at the University of Basel in 1588, where he earned an M.D. degree following submission of a thesis. It is not clear however, whether he traveled to Basel: his known activities at this time appear to me to make this unlikely.

[26]W. Pagel, review of *Die Alchemie,* in *Ambix* 13 (1966): 118–120.

characteristically displays a masterly knowledge of the bibliographical details of his subject, but he limits his commentary to an eclectic summary of some distinctive features of Libavius's description and preparation of chemical species.[27] Multhauf, in his perceptive analysis of the development of chemiatric pharmacology, notes the significance in Libavius's chemistry of preparations derived from the residues of distillation procedures, arguing that this marked an important turning point in sixteenth-century chemical pharmacy away from an almost exclusive concern with distilled fractions as remedies.[28] In these purely practical terms Multhauf sees Libavius as an ally of Croll. On a more theoretical level, Debus has examined Libavius's position in the debate over the legitimacy of Paracelsian medicine at the beginning of the seventeenth century, with particular reference to his contributions to the heated polemical debate which centered around the Paris Medical Faculty in the first decade of the century. Libavius is here depicted as one of those who sought a compromise between traditional Galenic theory in medicine and the innovative therapeutics of the chemical sect. As Debus demonstrates effectively, Libavius takes his place alongside others such as Guinther of Andernach and Daniel Sennert, who, although they rejected the cosmological and magical basis of Paracelsian theory, were significantly appreciative of the contributions made by chemical prescription to the enhancement of pharmacology that they sought to give an honored place to chemistry in materia medica.[29]

While I do not dispute any of the above conclusions, I do not believe that Libavius's goals for chemistry can be contained within the context of pharmacology or the purely medical disputes of the early seventeenth century. In what follows, I shall attempt to locate his concern for chemistry in a broader intellectual context, examining more closely his opposition to Paracelsianism, especially of the Crollian type, and to explicate the principles which governed his attempted rescue of chemistry from the hands of the followers of Paracelsus. Libavius has in fact provided us with ample documentation for this enquiry. The letter extensively cited at the beginning of this chapter is the first in a three-volume series of published epistles, of which the first two appeared in 1595 and the third in 1599.[30] These represent a

[27]Partington, *A History of Chemistry*, 2: 240–77.

[28]R. Multhauf, "The Significance of Distillation in Renaissance Medical Chemistry," *Bull. Hist. Med.* 30 (1956): 327–45, esp. pp. 343–45.

[29]A. G. Debus, "Guintherius, Libavius and Sennert: The Chemical Compromise in Early Modern Medicine," in *Science, Medicine and Society in the Renaissance: Essays to honor Walter Pagel*, ed. A. G. Debus, 2 vols. (New York: Neale Watson Academic Publications, 1972), 1: 151–65.

[30]See n. 1, above.

commentary on the principles and subject matter of the *Alchemia*, published in 1597. The first book of letters in particular provides important insights into Libavius's approach to writing his textbook. In addition, Libavius continued to write extensive commentaries on both editions of his textbook right up to the year before his death in 1616. The whole provides us with an extraordinarily excursive treatment of every aspect of chemistry as Libavius conceived it.

On the Dignity of the True Chemistry

Following his excoriating attack on the iniquities of the Hermetic-Paracelsian brand of chemistry, Libavius attempts to restore the balance in a second letter to his young friend by demonstrating that chemistry is indeed a legitimate form of human activity worthy of pursuit for its own sake.[31] As with the first letter, Libavius does not grapple with any intellectual objections to Paracelsian theory—he does not, for instance, try to establish alternative epistemological or cosmological principles for a reformed chemistry—rather he contrasts the antics of the Paracelsians with what he believes should be the behavior and social attitude of the genuine chemist. The letter reveals not so much a set of theoretical principles which should govern chemistry, as a set of attitudes which should inform its practice. The theme is that Paracelsianism represents an aberration of a genuine mode of enquiry and of a useful and worthy art. Chemistry requires, according to Libavius, intellectually and socially responsible practitioners. The expressed object of the letter is to remove the stigma attached to chemistry by demonstrating that the objections raised in the first letter are only applicable to Paracelsian abuses and not to the performance and objectives of the true art. In its debasement at the hands of the Paracelsians, chemistry has suffered no differently from other sciences and arts, such as theology, astronomy, law, and medicine, which have also been corrupted and abused by unscrupulous practitioners for their own gain. The integrity of any given art is not diminished or invalidated by particular and corrupt manifestations of it; it must be judged according to its essence.[32] Such an evaluation is one which only

[31]"De Verae Chymiae Honore," in *Rerum Chymicarum Epistolica*, 1: 19–47.

[32]"Una quidem est ars ipsa perpetuo secundum essentiam, sed non unus eius apud quoslibet habitus colorque, quemadmodum omnibus penè magnis euenit

learned men can make, and this, in Libavius's opinion, negates the objection made against chemistry that it has been publicly condemned and held in contempt. In this respect chemistry has suffered as much at the hands of its enthusiasts as it has at the hands of its detractors; for it is only through a scholarly appreciation and presentation of the art that a true estimate of its worth can be made. To this end, the art must be rendered capable of being judged in the ongoing dialogue of the scholarly world, and this in turn requires that the chemist conform to the accepted norms of scholarly behavior.[33]

Thus the true chemist will not shut himself up in his laboratory as the Paracelsians do, but when he has made a discovery, he will bring it out into the open and actively seek the scrutiny and approval of his peers for his work.[34] Nor will he keep others out of his laboratory. It is only by a rejection of Paracelsian secretiveness and arrogance that the accomplishments of the chemical art can be objectively assessed and put to public use. If he modestly sets out his discoveries from amongst the hidden glories of God in nature for public approval and acceptance, the true chemist cannot be accused, like the Paracelsian, of a total rejection of his responsibilities to God and society.[35]

Libavius next confronts the issue of the relationship of art to nature in an effort to remove both the philosophical and social stigma associated with the practice of the arts. He attempts to steer a narrow course between traditional Aristotelian reverence for the inherent perfection of nature as the object most worthy of man's study and the Paracelsian elevation of the crafts to a religious significance as a revelation and control of the divine powers in nature. Libavius, too, appeals ultimately to a religious sanction for man's arts, but one which remains purely on an ethical level. Far from representing a calumny against God's creation, man's arts are divinely sanctioned and are necessary for the health and well-being of society.

Libavius rejects the Paracelsian contention that nature provides nothing to hand for man's welfare.[36] Nevertheless, he does acknowledge that the chemist's primary concern is with the inner virtues and

---

artibus, veluti Theologiae, Astronomiae, Physicae, Iurisprudentiae, Medicinae & reliquis" (ibid., p. 20).

[33]Ibid., pp. 22–23.

[34]"Absolutum opus ubi est, non id abscondit, sed promit ad usus publicos; rationem miraculorum exponit, itaque locat, ut satisfactum sibi putet humana consuetudo. . . . Itaque ex officio absque pudore bona sua communicat chymiae professor, & iudices peritos ferre potest" (ibid., p. 24).

[35]Ibid., pp. 24–25.

[36]Ibid., pp. 26–27.

essences of natural objects, which cannot be brought to use except by art.[37] He effectively recognizes three levels of perfection in nature. First, there is the perfection of the species of creation, taken individually and collectively, which arouses study and admiration.[38] But it must be recognized that individual species do not always attain their final goal or the most perfect form of their being: the vagaries of climate and season and the recalcitrance of matter can hinder the full development of natural species. Here the art of man can enter in and assist nature in the realization of its end. Libavius specifically cites the activities of the farmer and the herdsman as examples of human skills being put to use to bring the products of nature, crops and animals, to a more perfect form than they would otherwise have attained.[39] Here man and nature actively combine to enhance the realization of a second level of perfection attainable in nature. But there is a third level of perfection in nature, attained principally by man's crafts. This involves bringing the end products of nature into a suitable state for man's use and consumption. Domestic crops and animals may possess their perfect natural form, but they still require cooking to be of use to man. With respect to nature they are perfect; with respect to the needs of man they are imperfect.[40] Thus a third perfection, defined and realized in terms of human needs and skills, can be drawn out of nature. Man in this respect transcends nature, and his ends are not necessarily coincident with the final ends of nature. This dominion of man over nature requires sanction outside of nature itself, and Libavius finds it in the Scriptures. It was God who authorized and commanded this mastery of man over nature and, according to Libavius, legitimized it by His example. God made coats of skins and clothed Adam and Eve on their expulsion from the Garden (Gen. 3:21); and He gave Moses the law on tablets of stone sculpted with His own finger (Ex. 31:18). These represented for Libavius divine sanctions of man's arts.[41] God gave

[37]"Scit [i.e., Chymicus] puras essentias propter varias coeli terraeq́; iniurias & incommoda non potuisse sine adminiculis nasci . . . " (ibid., p. 27).

[38]"Sua est cuiq; perfectio, quae naturae integritatē ad contēplandum foris cōmendat exornatq;" (ibid.).

[39]"Quicquid id sit, ars tentat amoliri, erratáque corrigere. Hoc cum agricolae & animalium curatores praestēt magna cum laude . . . " (ibid., p. 28).

[40]"Sed & alia perfectio naturae in se est, alia ad usum in penuria commodarum rerum. Sat perfectum forma naturali olus est; sat perfecta bovis caro: At ut bene nutriat, ab arte accipit commoditatem" (ibid., pp. 28–29).

[41]"Post peccatum Deus ipse non integra ove amiciebat Adamum, sed detracta pelle, &, ut credi par est, longè artificiosius divina arte elaborata quam fieri à pellionibus nostris fortè potest. Ita in exaranda lege non rudi utebatur lapide, sed affabrè per digitum Dei exculpto" (ibid., p. 29).

man the arts and also freedom of action, and if He wished to favor the human race in this way, who shall decry it?[42]

The character and not the force of Libavius's argument is what is compelling here. He is seeking a rationale for the intervention of man in nature on which he can base his claim for the dignity of chemistry. To do so he must argue around or supplant certain received opinions about the relationship of art to nature in conventional Aristotelian wisdom. Principal amongst these is that nature is the norm of art; human artifice is a mere imitation of the transformations and generations of the great artificer, nature. As such, art could not command the primary attention of the natural philosopher. There is the further problem of the extent to which art might represent an illicit intervention and a frustration of the processes of nature. "Chemistry destroys nature" was the charge which launched Libavius on this whole discussion. And lurking in the background is the sentiment that art can only transcend nature by means of magic. These are the problems uppermost in Libavius's mind in the course of this passage. They may strike us as remotely academic, and they probably were, even for many of Libavius's contemporaries for whom the artisan and the philosopher operated within completely different spheres. But they involved a critical issue for Libavius and kindred spirits, who were seeking to define new areas of human knowledge based on practical forms of knowledge and employing artisanal skills. Historical research increasingly reveals the prevalence and force of Libavius's assertion that European man's Christian heritage in the Bible—as opposed to his Greek philosophical heritage—gave him his cue for the systematic employment of his practical skills in an effort to master nature and make it serve his physical needs. The most noted and best studied exponent of just such an argument is Libavius's contemporary Francis Bacon. Farrington, Rossi, and Hooykaas have all stressed the scriptural foundation of Bacon's *Instauratio Magna.*[43] Libavius, it should be added, is, in this aspect of his thought at least, a more conservative philosopher than Bacon. He remains much more closely tied to his Aristotelian heritage and its respect for natural forms. Nature retains for Libavius its inherent

[42]"Neque huius poenitere creatorem potest, ut qui homines artibus donavit & libera agendi voluntate instruxit. . . . Si honorare humanum genus isto modo voluit, quis invideat?" (ibid., p. 31).

[43]See B. Farrington, *The Philosophy of Francis Bacon* (Chicago: The University of Chicago Press, 1966), esp. chaps. 3 and 4; P. Rossi, *Philosophy, Technology and the Arts in the Early Modern Era* (New York and London: Harper and Row, 1970), esp. appendixes 1 and 2; and R. Hooykaas, *Religion and the Rise of Modern Science* (Edinburgh and London: Scottish Academic Press, 1972), esp. pp. 54–74.

perfection of forms appropriate for our contemplation. His view of the agriculturalist as assisting nature to achieve its final forms is also wholly orthodox. But he shares with Bacon the notion that there is a greater good outside of nature itself which is man's well-being and welfare. Man's ends have higher sanction than nature's teleological goals. The argument was far from irrelevant and trivial: it was on just such a basis that the scientific and technological exploitation of nature was launched in early modern Europe.

While Libavius offers a biblical justification for man's dominion over nature, he does not attach any inherent spiritual significance to the arts, as do Paracelsus and Croll. For Paracelsus and Croll, man's alchemical crafts involved manipulation of divine powers in nature; the artist participated in a divine-cosmic process and was the agent through which the divine in nature was made manifest. The Paracelsian world view was predicated on the belief of the immanence of divine powers in man and in nature, which broke down all barriers between the natural, human, and divine. Libavius resists all such conflation of knowledge and power: his is still a hierarchical world in which nature, man, and God have their own appropriate spheres and modes of operation. Man does not mediate, legitimately at least, between the divine and the natural. He has power over nature by means of his arts, but it is not an unlimited power; it is constrained by the inherent powers of nature, which man can channel to his own purposes but which he cannot contravene or negate. Only God has the power to suspend the normal powers of nature and produce miracles. Man has divinely sanctioned authority to employ nature for his own legitimate ends, but he is not a magus who can marry heaven to earth.

It is with this much more circumspect view of man's arts that Libavius attacks the pretensions of the Paracelsian chemistry. He holds the Aristotelian division of the world into sublunary and supralunary spheres sacrosanct, and he denies the Paracelsian claims to preparing the quintessence free from all material and elementary composition. It is simply not within the compass of human art to prepare the true fifth essence, and the Paracelsians are guilty of fraud when they sell their preparations as such.[44] The true chemist does not disguise his elementary preparations but offers them for what they are—oils, waters, and the like.[45] The object of chemistry is not to destroy the material matrix of natural substances; it is directed at producing individual discrete preparations. Thus the true chemist employs only the appropriate degree of fire to liberate the interior substance which he wishes to derive from

[44] *Rerum Chymicarum Epistolica,* 1: 28.
[45] Ibid.

the more solid material matrix which encloses it; his aim is not to destroy the *mixis.*[46] The transmutations which the chemist can perform are determined and limited by the motions and powers inherent in nature.[47] Man cannot produce instantaneous and miraculous transformations in natural objects; such power resides in God alone. The Paracelsians lay claim to such capabilities, but then they impiously profess magical powers and do not abstain from such stupidity.[48] Libavius's argument amounts to a wholesale attack on the concept of man as a magus. Man cannot legitimately lay claim to divine powers in magic. The power he exercises over nature comes from a cooperation with the forces within nature, which he can direct to his own ends.

Although Libavius would appear here to be setting forth a conventional medieval view of man, recognizing both his natural and spiritual limitations, in contradistinction to the Renaissance concept of the divine man, who bridged heaven and earth, time and eternity, he was not uncritically bound to a respect for received wisdom. The dynamic element in his thinking is provided by his concern for the development of knowledge for human use. It is an enterprise which he recognizes will not be accomplished on an individual basis: it will require collective effort; and within this cooperative enterprise there will be need for specialization. To illustrate his point he cites St. Paul's description of the division of ministries within the Church.[49] There is an urgency in his appeal. He deprecates the time given over to "heavenly" theory and the attitude of mind which demands perfection in this imperfect world. It is sufficient, he states, that we investigate properly those things of which we are "shamefully" ignorant for human use.[50] In this respect the writings of the ancients are of limited use, firstly, because the men of antiquity did not hold the application of art in high regard, and secondly, because the passage of time has reduced the value of their testimony. There have been so many additions to the arts and so many variations in practice through time and according to place and ability that the only satisfactory procedure is to follow contempo-

[46]Ibid., pp. 30–31.

[47]Ibid., p. 32.

[48]Ibid.

[49]"Sed neque hominis ea provincia est. Multa eáque dissentanea rectius aguntur à multis. Itaq; & in ecclesia & officia & personae sunt à divo Paulo distinctae" (ibid., pp. 32–33).

[50]"Multitudo & copia non est in mora. Satis est si quae turpiter ad usum humanae vitae ignorantur, investigantur probè. Maxima pars ad Theoriam coelestam est reiicienda. Qui in imperfecto vivendi & potētiarum manco ministerio perfectionem exigit, is se parum aequum rerum iudicem esse declarat" (ibid., p. 33).

raneous practice and usage. By following this pragmatic advice, Libavius suggests, if one does not follow how the artist arrived at his goal, at least one will not be ignorant of the method by which he sought to achieve it.[51] He describes as superstitious the debates of the philologists on the details of punctuation in the writings of antiquity carried on in a spirit which would imply the denial of any possibility of error to "ancient" man.[52] This is the obverse of the extreme exemplified by the Paracelsians, who deny discovery of any sort to the ancients. Libavius here as elsewhere displays a basic conviction in the fallability of human knowledge. Wisdom is not the monopoly of the ancients, nor of the newborn magi; it is the product of a collective and cumulative process which emerges in the course of informed and scholarly dialectic through historical time.

The most remarkable aspect of Libavius's true chemistry (*chymia vera*), however, was the vision he entertained of its independence and integrity as a discipline. He was not, in 1595, describing a subject which already existed, but one which he was trying to create. In the course of this second epistle to his young friend, he forcibly argues for an independent chemistry which was not justified solely in terms of the services it could render to medicine. He asserts that it cannot be denied that chemistry confers many benefits on the medical art and that the physician who can practice chemistry is furnished with a notable addition to his professional skills.[53] But being a chemist does not make one a physician. If the chemist effects a cure, he does so, not on the basis of the principles of chemistry, but according to principles of the healing art, as set down by Hippocrates and Galen.[54] The two disciplines are distinct; historically medicine has done without chemistry, and it can continue to do so.[55] He makes an impassioned appeal for the independent dignity of chemistry:

[51]"Veterum dicta explanare illis difficile est, qui artis usum non fecerunt laudatum, quanquam dandum aliquid sit iniuria temporum, quibus historica fides est labefactata. Artium sunt incrementa mirifica; varietàtes etiam per tempora, regiones & ingenia multae. . . . Remedio est praesentis praecepti secundum praesentem consuetudinem inculcatio, ut si non assequaris quo illi modo ad finem pervenerunt, saltem non ignores quo commodè iam queat perveniri" (ibid.).

[52]"Omnia puncta & apices literarum solicitè disquirere, superstitiosi est, & homini vetusto omnem errandi potentiā cum magna vesania detrahentis" (ibid., p. 34).

[53]Ibid., p. 38.

[54]Ibid., p. 40.

[55]"Caruit & carere potest medicina chymico apparatu quod artis essentia attinet" (ibid., p. 41).

> But when I say that chemistry ought to be of assistance to medicine, do not think that it is so servile an art as to have no domain of its own nor any luster and distinction peculiar to itself. If serving and ministering were to render a science contemptible, which one of them would enjoy any dignity or honor since all serve one another in turn? The chemist does not always or only stand in wait. But [chemistry] can hold its head as high as any branch of philosophical enquiry has ever done. It rejoices first of all in its own perfection and even when its works serve other [arts], its primary and proper end is not servitude, but its own prosperous independence.[56]

Libavius asks why such self-justifying status cannot be granted to chemistry when it is already enjoyed by physics, which teaches nothing more after it has duly described nature. In support he cites the many discoveries he personally has made about the workings of nature through chemistry, which have allowed him to understand the powers of action and passion which reside in natural objects. He drives home his point in a condescending comparison with contemporaneous physics. Many, he asserts, dispute about the vacuum, infinity, motion, and the eternity of the world, but when they have anxiously unraveled as much as possible about these separate topics, they have learned no more than the average educated person already knew. The chemist, on the other hand, possesses a precise and certain knowledge of the sympathies, antipathies, and causes of things, having investigated their effects.[57] But the great virtue of the chemical artist is that he produces not just isolated speculation, but a useful and beneficial praxis for society, appreciating that no art is worth its salt which does not produce practical consequences from its theory.[58] Libavius concludes that chemistry is a genuine part of physical investigation (*physicae contemplationis*), but one which goes beyond this to contribute to the sacred art of medicine.[59] It is at this point that he utters one of his unforgettable excoriations against the Paracelsians: "Paracelsianism will be phil-

[56] "Sed cū medicinae iuvandae esse chymian dico, ne putes tam servilē esse artem, ut nihil habeat dominij nec quicquā splendoris proprij & magnificentiae. Si servitium & ministerium contemtam redderet scientiam, cum omnes sibi mutuo serviant, quaenam aliqua dignitate & honore frui posset? Non semper nec tantum ministrat chymicus. Sed caput exerit tam altè, quam ulla philosophicarum contemplationum exeruit unquam. Ipsa in sua perfectione acquiescit primùm & cum eius opera inserviant aliis, primus tamen finis & proprius servitium non est sed beata quaedam αυταρχεια" (ibid., pp. 38–39).

[57] Ibid., pp. 39–40.

[58] "Sciūt enim artifices chymici non tam prodesse societati humanae solitariam speculationē, quam cum ea coniunctam praxin usumq; salutarem, & nullam artem divinam excellentemque esse, quae non actionem Theoriae habeat cōsequentem" (ibid., pp. 41–42).

[59] "Omnino enim vera chymia pars physicae contemplationis est, quae postea cum sacra conspirat medicina, sicut aliae multae" (ibid., pp. 43–44).

osophical when all the whores are chaste virgins and all sophisms are indubitable truth."[60]

The above passages of Libavius would not be out of place in the early nineteenth century, but they seem in retrospect an astonishing statement for 1595. Even more perplexing is their source and objective—a crusty Lutheran academic seeking to set more noble goals for a wayward former pupil who is being seduced by Paracelsianism. It would appear to reveal a greater resourcefulness than seems likely from within the humanist educational tradition of Lutheranism. Libavius's humanism, however, was broad enough to countenance the benefits which the chemical arts offered to society, and he was prepared to bring them within the scholarly ambit. Our conservative schoolmaster turns out to be a remarkable man indeed—he seeks to combat the counterculture by coopting it.

60"Paracelsica professio tunc philosophica erit, cum scorta virgines honestae & sophismata ipsa veritas indubia" (ibid., pp. 44).

CHAPTER V

THE CLASSROOM OF THE MIND

In 1615, the year before his death, Libavius published the last of his great folio volumes, which served as commentaries to the second edition of his *Alchemia.*[1] The bulk of the work, entitled *Appendix necessaria syntagmatis arcanorum chymicorum,* is in fact given over, not to his own chemistry, but to a series of intensely polemical pieces written in response to personal opponents and a variety of recently published treatises advocating Hermetic and Paracelsian views. It is Libavius's last appeal for a rational and responsible chemistry and his final attempt to stem the apparently swelling tide of enthusiasm for the irrational and magical Hermeticism and Paracelsianism. Characteristically, we find Libavius doing battle on two fronts. The second tractate in the series is a defense of alchemy, written in reply to a personal attack made against Libavius in an anti-alchemical work published in 1614 by the Lorraine physician Nicolas Guibert. This was the last salvo in a dispute between the two antagonists over the possibility of transmutation which had begun in 1603.[2] The remaining polemical sections are directed against various manifestos for Hermetic

[1]*D.O.M.A. Appendix necessaria Syntagmatis Arcanorum Chymicorum Andreae Libavii M.D.P.C. Halli-Saxonis illustris Gymnasii apud Corburgenses Directoris, Professoris publ. et Medici Chymici-practici. In Qua Praeter Arcanorum Nonnullorum expositionem & illustrationem, quorundam item Medicorum Hermeticorum, & mysteriorum descriptionem, continentur defensiones genuinae . . .* (Frankfort, 1615).

[2]Nicolas Guibert (c. 1547–1620) was in his youth a widely-traveled alchemist who came to repent the error of his ways and published a refutation of the art in *Alchymia Ratione et Experientia ita demum viriliter impugnata & expugnata* (Strasbourg, 1603). This evoked a vigorous response from Libavius in his *Defensio et Declaratio Perspicua Alchymiae Transmutatoriae, opposita Nicolai Guiperti* (Ursel, 1604). The dispute was resumed in Guibert's *De Interitu Alchymiae Metallorum Transmutatoria Tractatus aliquot: Adiuncta est eiusdem Apologia in Sophistam Libavium, Alchymiae refutatae furentem calumniatorem* (Toul, 1614), to which the tract under discussion in the present volume was Libavius's reply. For further details on Guibert see Martin Fichman, "Guibert, Nicolas," in *D.S.B.*, 5: 579–80; J. R. Partington, *A History of Chemistry*, 4 vols.

and Paracelsian medicine and philosophy. The best studied of these is the one devoted to a critique of the Rosicrucian manifesto and the *Fama Fraternitas,* which has recently brought Libavius to the attention of Frances Yates.[3] Defenders of Paracelsus who draw the ire of Libavius include Henning Scheunemann, author of a number of works on Paracelsian theories of disease,[4] and Johannes Hartmann, professor of chemiatry at the University of Marburg.[5] But the polemic to which I wish to draw attention here is that directed against Oswald Croll, prompted by the publication of the latter's *Basilica Chymica* in 1609.[6] This polemic, which provides Libavius's most detailed critique of the whole theoretical and epistemological basis of Paracelsian chemistry, throws into relief the fundamental differences in world view which separated two important systematists of chemistry at the beginning of the seventeenth century. Whereas in his chemical epistles which preceded the publication of the *Alchemia* Libavius confined his criticisms of the Paracelsians to an attack on their conduct, here he takes issue directly with the intellectual beliefs and doctrines which underlay that behavior. Although this critique of Croll postdates Libavius's synthesis of the chemical art in the *Alchemia* by some eighteen years, it is appropriate to consider it at this point, since it provides the most succinct statement of the latter's basic epistemological objections to Paracelsianism, which had prompted his reformulation of chemistry.[7]

(London: Macmillan, 1961–70), 2: 268; and Lynn Thorndike, *A History of Magic and Experimental Science,* 8 vols. (New York: Macmillan, 1923 [vols. 1–2]; Columbia University Press, 1934–58 [vols. 3–8]), 5: 648, and 6: 244–47, 451–52.

[3]Frances A. Yates, *The Rosicrucian Enlightenment* (London and Boston: Routledge & Kegan Paul, 1972), pp. 51–56 and passim.

[4]Henning Scheunemann published several works on Paracelsian theories of disease between 1608 and 1613. His magnum opus, however, did not appear until 1617 as *Medicina reformata seu Denarius Hermeticus. . .* (Frankfort, 1617). See Thorndike, *A History of Magic and Experimental Science,* 7: 174; and Partington, *A History of Chemistry,* 2: 250.

[5]Johannes Hartmann (1568–1631) was appointed professor of chemiatry at the University of Marburg in 1609—the first such appointment in a European university. He was a protégé of the Landgrave Moritz of Hesse-Kassel, yet another princely devotee of the occult and alchemical sciences. Hartmann used Croll's *Basilica Chymica* in his practical courses at Marburg. For further details see R. Schmitz, "Hartmann, Johannes," in *D.S.B.*, 6: 145–46.

[6]The polemic against Croll, Hartmann, and the Rosicrucians contained in the *Appendix necessaria* has the separate title page: *D.O.M.A. Examen Philosophiae Novae, Quae Veteri Abrogandae opponitur: in quo de modo discendo novo: de veterum autoritate: de Magia Paracelsi ex Crollio: de philosophia vivente ex Severino per Johannem Hartmannum: de philosophia harmonica magica Fraternitatis de Rosea Cruce: Opera & Studio Andreae Libavii. . .* (Frankfort, 1615), hereinafter to be cited as *Examen Philosophiae Novae.* There is a general introductory preface on pp. 3–12, and the section on Croll occupies pp. 13–87.

[7]W. Hubicki in his article on Libavius in *D.S.B.*, 8: 310, alleges that Libavius's attitude to the Hermetic-Paracelsians changes dramatically in the period

At the outset it should be stated that although Libavius incorporates a defense of Aristotelian epistemology into his attack on the Paracelsians and Hermeticists, it is soon evident that the issue for him is not simply a philosophical one which could be settled exclusively in terms of a confrontation of epistemological doctrines. Wider issues are at stake. The conflation of divine wisdom and power with human wisdom and power in Croll's presentation of Paracelsianism confronted the orthodox Lutheran schoolmaster with particular problems which went beyond the range of the purely philosophical: it involved the criteria for religious truth in addition to those for purely secular knowledge. Of necessity this dictated a delicate strategy on the part of Libavius, for any Lutheran was very susceptible to being hoist with his own petard by the "enthusiastic" character of Croll's arguments. When Luther challenged the rule of faith of the Church as the canon of revealed religion and sought to substitute in its place the Word of God in the Scriptures subjectively interpreted in the light of grace, he opened the Pandora's box of Scepticism, which was to engulf both theology and philosophy in the sixteenth and seventeenth centuries. The employment of Sceptical and Pyrrhonist arguments by both Reformers and Counter-Reformers proved to be a dangerous double-edged sword which was to have far-reaching effects on European philosophy in the early modern period.[8] While Libavius's answer to the Paracelsian

1608–10. He notes that previous to this time Libavius had good personal relations with Croll, Hartmann, and the Landgrave Moritz, among others. Hubicki suggests two reasons for this transformation: (1) Libavius's increasing religious orthodoxy and (2) his disappointment at not being awarded the chair of chemiatry at Marburg. This, I believe personalizes the issue too much, particularly in instance (2). As is clear from Libavius's chemical epistles of 1595, discussed in the previous chapter, the Lutheran schoolmaster was an implacable foe of Hermetic Paracelsianism from that early date. This did not, however, prevent him from having friendly contacts with advocates of this philosophy on account of their chemical knowledge and interest. What changed this, I believe, was the emergence in just this period (1608–10) of what Libavius perceived as an unholy alliance between Calvinism and Hermetic-Paracelsianism. For in the same year as Croll's *Basilica Chymica* (1609) the Landgrave Moritz of Hesse-Kassel, having earlier converted to Calvinism and expelled the Lutherans from the University of Marburg, the first Lutheran University of Germany, installed Hartmann in the chair of chemiatry. If one adds to that the appearance of the Rosicrucian manifestos from the Lutheran center of Tübingen in 1614 and 1615, one has adequate explanation of Libavius's polemics of 1615. He surely felt he was witnessing the fall of the intellectual centers of Lutheranism to the increasingly allied evils of Calvinism and Paracelsianism. While these developments are significant, they do not affect our present analysis, however. We shall be concerned principally with the epistemological issues between Croll and Libavius, and here the latter's position is perfectly consistent with his statements prior to the publication of his *Alchemia* in 1597.

[8] For a discussion of this issue, see Richard H. Popkin, *The History of Scepticism from Erasmus to Descartes* (New York: Harper Torchbooks, 1968), esp. chap. 1, "The Intellectual Crisis of the Reformation," pp. 1–16.

theory of knowledge cannot be set exclusively in the context of the Sceptical debate, some of his problems stemmed from the same issues which gave rise to that conflict, and it is significant that in one of his sallies against Croll he attempts to drive the Paracelsians into the Pyrrhonist camp.

Libavius's dilemma in the face of Croll's agruments can be plainly stated. Croll's assertion that the unique source of all wisdom was the Word of God in Scriptures interpreted in the light of grace could not be casually dismissed by a pious Lutheran. At face value, it was an orthodox statement of the Lutheran rule of faith, and to reject it out of hand in favor of a restatement of a rational epistemology based on Aristotle was scarcely an adequate response under the circumstances. Libavius had first to divorce the religious issue from the philosophical one and then to attack each separately. This is how he proceeds in his treatise against Croll. The first section is devoted almost exclusively to the epistemological issues involved; the succeeding three sections attempt to demonstrate that the magical and Cabalistic interpretation by Croll of the Word of God is both impious and at variance with the Scriptures themselves.[9] But even this tactic does not wholly solve the problem, for to make overly confident claims for rational knowledge on the basis of Aristotle offended Lutheran piety, which stressed the baseness and helplessness of man as a result of the Fall. If too strong claims were made for the ability of unaided reason to achieve certain knowledge, a path was opened to a rational metaphysic and a natural theology, a scholastic specter which was anathema to Lutheranism. A divorce of theology and philosophy could not be had simply by containing the two within separate sections of a book. Libavius finds his way round these delicate problems, not surprisingly, by drawing from the wellspring of Lutheran intellectualism, the humanist tradition of Melanchthon.[10] According to this, knowledge could be gained by

[9]The polemic against Croll (*Examen Novae Philosophiae*) is divided into four separate disquisitions as follows: (1) *D.O.M.A. Ex Praefatione Oswaldi Crollij in suam Basilicam Chymicam. De Cognitione Philosophiae, et Medicinae, Prima Disquisitio*, pp. 13–22; (2) *D.O.M.A. Positne Paracelsici Magiam Suam Divinarum Litterarum Autoritate, Et Philosophiae, Physicaeque Nomine Defendere*, pp. 22–34; (3) *D.O.M.A. Paracelsica Sententiarum Biblicarum Depravatio ex Oswaldi Crollii Praefatione Admonitoria, Exemplo dicti Paulini I. Thessal, cap. 5, v. 23*, pp. 35–60; and (4) *Exercitatio Alia De Abominabili Impietate Magiae Paracelsicae per Oswaldum Crollium Aucta*, pp. 61–87.

[10]The classic study of Melanchthon as the fountainhead of Lutheran education and intellectualism remains K. Hartfelder, *Philipp Melanchthon als Praeceptor Germaniae* (Berlin: A. Hofmann, 1889). See also P. Petersen, *Geschichte der Aristotelischen Philosophie im' Protestischen Deutschland* (Leipzig: Verlag von Felix Meiner, 1921). For a brief but brilliant analysis of Melanchthon as an intellectual see Franz Hildebrandt, *Melanchthon: Alien or Ally?* (1946; reprint ed.,

sense experience and reason, but only a rash individual indeed would claim absolute conviction for naturally acquired knowledge, given the impaired status of man's faculties since the Fall. Aristotle, (and in Melanchthon's case also Galen),[11] provided the ground rules for the attainment of natural knowledge, but these rules had to be put into operation by fallible men, who were liable to err both in the sensory evidence they adduced and in the rational deductions they made from that evidence. Progress in human knowledge (*scientia*) had therefore to be a collective endeavor in which the contributions of the individual were subject to the scrutiny of his compeers and had also to be measured against the collective wisdom of past ages. This cumulative and collective character of human knowledge threw into prominence the communicative aspects of the interpretation and exposition of knowledge. The tools for this task lay in the beloved trivium of the humanists, grammar, dialectic, and rhetoric.[12] Libavius emerges as an opponent of the Paracelsian theory of knowledge, not primarily as a philosopher *qua* epistemologist but more significantly as a humanist *qua* dialectician. It is the skills of the trivium, that is, the skills imparted by the schoolmaster, which he proposes as the saviors of learning from the twin perils of individual enthusiasm and wholesale Pyrrhonism. Libavius's quarry, as might be expected given his background, turns out to be intellectual stability, not intellectual certainty.

In his rebuttal to Croll, Libavius adopts a pincer strategy of attack. On the one hand, if the Paracelsian epistemology is taken seriously, the inevitable result is intellectual chaos. On the other hand, if it can be shown that many Paracelsian teachings derive in fact from earlier philosophies, then Paracelsianism is not the radically new philosophy its proponents claim. Rather it becomes part of the overall European intellectual tradition, and as such, its doctrines must be assessed in the light of the collective wisdom of that tradition and

---

New York: Kraus Reprint Co., 1968). Two articles by Quirinus Breen are of special relevance to the present study: "The Subordination of Philosophy to Rhetoric in Melanchthon. A Study of his Reply to G. Pico della Mirandola," *Archiv für Reformationsgeschichte* 43 (1952): 13–28; and "Giovanni Pico della Mirandola on the Conflict of Philosophy and Rhetoric; Melanchthon's Reply to G. Pico della Mirandola," *J. Hist. of Ideas* 13 (1952): 384–426.

[11]For a detailed study of Melanchthon's theory of knowledge see J. Kump, *Melanchthon's Psychologie (seine Schrift de anima) in ihrer Abhänigkeit von Artistoteles und Galenos,* Inaugural dissertation of the philosophy faculty of the University of Jena (Kiel, 1897).

[12]Melanchthon divided philosophy into three parts: (1) the *artes dicendi* (dialectic and rhetoric); (2) *physiologia* (epistemology, physics, and mathematics), and (3) *praecepta de civilibus moribus* (ethics). See Kump, *Melanchthon's Psychologie,* p. 3. The summation of all philosophy was moral and civic responsibility. Cf. Breen, "The Subordination of Philosophy to Rhetoric," p. 27, and Libavius's attitudes discussed in the previous chapter.

judged by its generally accepted canons of criticism. It is important to recognize the essentially humanist character of Libavius's critique because as a purely philosophical exercise it is less than overwhelming.

In the general preface of the *Appendix necessaria syntagmatis arcanorum chymicorum,* which serves as an introduction to his polemics against Croll, Hartmann, and the Rosicrucian Fraternity, Libavius sets out the elements of his strategy. In noting the wholesale attack on the philosophy of the Gentile ancients mounted "in this last age," Libavius cautions that careful consideration must be given to the nature of the refutation and condemnation, "lest we seem to defraud ourselves and our pupils with lies instead of truth."[13] The pedagogical concern is wholly characteristic: the philosophical issues are secondary to the maintenance of sound doctrine in the schools. The main areas of debate, contends Libavius, are in philosophy and medicine; for although there has been some contention about the *prisca theologia,* this should not cause difficulty for those who believe God and not men, especially false (*non authenticis*) interpreters of the Scriptures.[14] This assertion begs many theological questions—for instance, who is an authentic interpreter of the Scriptures?—but it serves for the moment to eliminate certain religious matters from the discussion.

Libavius points out that Aristotle has borne the brunt of the assault. He cites the attacks made on the latter's philosophy by Bernardino Telesio, Peter Ramus, and James Martin, a Scottish disciple of Ramus whom Libavius earlier had attacked on questions of natural philosophy (see chapter VI). But of all the innovators, none has been more admired than Paracelsus, who has attracted a great following. Libavius lists amongst this group Johannes Hartmann, professor of chemiatry at the University of Marburg; Oswald Croll; Henning Scheunemann, the Bamberg physician; Joachim Tachenius (1537–1609), a professor of anatomy at Leipzig; and the Fraternity of the Rosy Cross—all of whom were active in Germany during the first two decades of the seventeenth century.[15] Significantly, the list points up the fact that Libavius was not responding to a general philosophical trend but to a movement which was continuing to make important headway in German academic and medical circles.[16] This last group, alleges Libavius, wish to dispense with all the writings of earlier philosophers, as

13". . . diligenter considerandum est, cuius modi sit illa seu reprehensio, seu damnatio, ne pro veritate mendaciijs & circumducti nos videamur, & discipulos nostros circumducere" (*Examen Philosophiae Novae,* p. 3).

14". . . tamen de hac [i.e., antiqua theologia] minus dubitare possunt illi qui Deo potius esse credendum quam hominibus sciunt, atque ita facile, quae de interpretibus scripturarum non authenticis asseruntur, dissolvere" (ibid.).

15Ibid.

16See n. 7, above.

well as with Galenic medicine, which must cause one to speculate on the reasons behind the wholesale condemnation of the ancients, since one of their number, Croll, has written that "nothing can be said which has not been said before . . . there is nothing new under the sun."[17]

Libavius goes on to contend that the Paracelsians have not really rejected the heritage of antiquity: they have simply divided it into two camps. One is the magical Cabalistic and Hermetic tradition, which had its roots in Chaldea and Egypt and has recently been revived in Germany under the sponsorship of Paracelsus, whose followers have taken the name Theophrasteans to give themselves the aura of antiquity. The other is the Peripatetic, Galenic, and Arabic tradition, which includes Pliny, Dioscorides, Avicenna, and Mesue, all of whom are rejected by Paracelsus.[18] The dispute, therefore, is not between ancients and moderns but between two living historical traditions. Within the scope of a page Libavius, not inaccurately, has put the dispute in a context in which he can handle it more effectively.

The problem remains, however, of establishing a dialogue between these two traditions. Libavius notes that the only response of the Paracelsians to earlier criticisms of their philosophy has been vituperative ranting.[19] They do not reason from first principles or even demonstrable facts, but from their own opinions and phantasies concerning the relationship between the great and the little world. They explicitly reject the writings of the ancients, especially those they regard as pagan, and instead put their trust in the light of nature and special revelation by the light of grace.[20] Libavius counters by citing a favorite text of Christian humanists, from Paul's Epistle to the Romans (Rom. 1:19, 20): "For what can be known about God is plain to them [i.e., the Greeks] because God has shown it to them. Ever since the creation of the world his invisible nature, namely, his eternal power and deity, has been clearly perceived in the things that have been made." The Gentile Greeks thus had access to a knowledge of God and his creation through the light of nature. The Jews, on the other hand, were the recipients of

[17] *Examen Philosophiae Novae* p. 3.

[18] "Quia vero antiquitatem in duas factiones dirimunt, & unam Magicam, Cabalisticam, Hermeticam, &c. ex Chaldaea & Aegypto, & nuper ex Germania, Paracelso autore, unde Theophrastaea Paracelsica dicitur, quasi & haec sit vetusta, faciunt: alteram Peripateticam Galenicam Arabicam &c., ita ut simul Plinium, Dioscoridem, Avicennam, Mesuen, & quoscunq[ue] alios, quos omnes in tergum suum rejicit Paracelsus, comprehendant, & nonnullis utantur argumentis, operae precium erit quibus fundamentis videre" (ibid.).

[19] Ibid., p. 4. Three critics of Paracelsus whom Libavius specifically mentions in this section are Thomas Erastus, Johann Crato of Krafftheim, and Jacob Zwinger of Basel.

[20] Ibid.

a knowledge of God through the light of grace. In Christianity the two lights are joined, and the philosophy of the pagan ancients derived from the light of nature has been examined, illuminated, and fulfilled by the divine revelation.[21] The two lights are thus historical traditions contained within their respective written memorials, the writings of the ancient philosophers and the revealed Word of God in the Scriptures. This is the classic defense of Christian humanism, whose principal Lutheran spokesman was Melanchthon.[22]

It is this humanist interpretation of the two lights which Libavius places in opposition to the enthusiastic interpretation of the Paracelsians. In a passage which presents a succinct statement of his views, he denies the validity of unique claims to special illumination in either light:

> Then in the face of so many contrary opinions and differing judgments, it must be asked whose light of nature and grace is true and whence does it come, since it is agreed that the devil can simulate the angel of light and cast out images of truth instead of truth itself, as may be the case in the similitudes, analogies and harmonies of Paracelsus. Even Plato, who was familiar with the barbaric wisdom of Egyptians as well as the Greek wisdom of Homer and the philosophers, does not in any way remove the knowledge of truths from human dialogue. . . . Thirdly all the power of experience and of the scientific nature of disciplines [would] perish, since the life and ability of no one man is so great that he can know everything and confirm it by his own experience. Fourthly, if the light of nature is substituted in place of argument, it cannot be that nature which has been corrupted by a curse and made subject to vanity since the fall of the angels and of man; it can only be that nature which Adam uniquely possessed and which the Saints will possess in their state of perfection. Nor may one seek assistance towards perfection in this generation from the light of grace. For in this life that [light] is so particular, incomplete and imperfect as the Apostle writes in 1 Cor. 13, and as the facts themselves demonstrate. The carnal affections so dim the light of the mind, that even when illuminated it wavers, and still less should it be abandoned to its own resources. Those who claim for themselves the light of grace, display it in a wondrous fashion and present I know not what revelations. Consequently it remains for us to join our knowledge to that of the ancients, testing everything and holding only to that which is good.[23]

Claims to special revelation are not only absurd, they are also dangerous, as the imagination is subject to powers of evil as well as to the powers of good. In any event, man in his post-lapsarian condition cannot aspire to certain knowledge of the essences of things; the individual can only pool his intellectual resources in a spirit of humility

[21] Ibid.

[22] See Hildebrandt, *Melanchthon,* esp. pp. 1–33.

[23] *Examen Philosophiae Novae,* pp. 4–5.

with those of others to make his contribution to the attainment of as perfect a wisdom as is possible in this imperfect world. This demands intellectual discourse, as even Plato recognized in casting his philosophical works in the form of dialogues.

Libavius, however, does not espouse an eclectic position in philosophy. From amongst the schools of ancient philosophers a choice must be made based upon compatibility with Christianity. This was Melanchthon's position,[24] and Libavius makes the same choice. Of all the philosophies conceived in ignorance of the Church, the school which *disagrees least* is the the Peripatetic philosophy. Hence to philosophize in a Christian manner is to philosophize Peripatetically.[25] The claim of the Paracelsians to possess the unique Christian philosophy in their Cabalistic, magical, chemical system is quite fallacious, as no support can be found for it in the fount of Christian belief, the Scriptures. Their error is to confound the two sources of knowledge, the light of nature and the light of grace, in an all-encompassing enthusiastic interpretation which reaches the height of vanity and impiety, in their search for the Word incarnate in natural things.[26] The words of the Scriptures are the touchstone of knowledge about the Word incarnate; the words of the philosophers, principally the Peripatetics, are the touchstone of knowledge about nature.[27]

In his philosophical analysis of Croll's statement of Paracelsian epistemology, Libavius quite clearly lacks the tools to provide an effective critique. He can only fall back on Aristotelian categories of knowledge–boundaries between metaphysics and natural philosophy, the elements of Aristotelian faculty psychology, distinctions between essential knowledge and "accidental" knowledge, as well as the particular doctrines of Aristotle's natural philosophy–which prove wholly inadequate to digest and refute Paracelsian gnosticism, which equated knowledge with spiritual illumination and the harmonies of the great and little world. These latter beliefs completely shattered Aristotle's compartmentalization of human knowledge: they defied analysis in Peripatetic terms. Much of Libavius's response, therefore, necessarily devolves into an exploitation of what he considers inherent contradictions in Croll's statements, in an effort to break down the latter's

[24]Hildebrandt, *Melanchthon*, p. 6.

[25]*Examen Philosophiae Novae*, p. 6.

[26]Ibid., p. 7.

[27]"Ad haec claves quaedam requiruntur, ut linguae sapientū intellectus, interpretatio Doctorū, collatio, illuminatio spiritus, exercitatio crebra, quae parit habitum, & ut ante dictum est, post gratiam divinam Dialectica, Rhetorica, Grammatica. In sacris nosti à Dei mandatis, & verbo non esse discedendum, neque ad dextram, neque ad sinistram" (ibid., p. 16).

doctrines into elements which are subject to refutation by conventional Aristotelian arguments. His most frequent tactic is to attempt to identify Croll's epistemological views with those of some of the pre-Socratics whom Aristotle had criticized. This serves a dual purpose: first, it provides a ready-made counterargument, and second, it allows him to charge that the Paracelsian doctrines were not original or uniquely Christian in inspiration, but simply derivative from pre-Aristotelian, pagan philosophies.

Libavius claims an inconsistency in Croll's assertion that the source of all knowledge is God and his statement that man possesses within himself all knowledge of natural things.[28] But Libavius fails to set this in the context of Croll's distinction between the intellectual soul and the astral body and the different roles these play in cognition. In one place Libavius argues that if Croll's claim that the source of the invisible philosophy of the Paracelsians is the intellectual soul of man, "which is the divine light flowing out of the spiracle of God," then this philosophy will never pass beyond the bounds of metaphysics, since it will be limited to a contemplation of the soul by itself.[29] This is not a tack which he pursues to any great length, however. Rather he focuses on the doctrine of astral cognition, whereby knowledge is acquired through the sympathetic interaction of the astra in man the microcosm and the astra of the great world. Here Libavius detects an affinity with certain pre-Socratic doctrines which Aristotle refuted in the *Metaphysics* and in *De Anima.*

In the first instance Libavius identifies the Paracelsian assertion of the primacy of individual "experience" in the light of nature with Protagoras's doctrine that man is the measure of all things, that is, that reality is determined by individual opinion.[30] Aristotle's main discussion of the subjective relativism of Protagoras occurs in book 4 of the *Metaphysics.*[31] Here Protagoras is considered in the company of Democritus, Empedocles, Anaxagoras, Heraclitus, and Parmenides. The context is principally the failure of these philosophers to understand

[28]Ibid., p. 13.

[29]In response to Croll's contention that the invisible philosophy pertains to the intellectual mind of man Libavius writes: "Non enim si anima hominis est metaphysicae per se contemplationis, nihil praeter metaphysicam docendum est in Philosophia, cuius non id tantum est subjectum & materia . . ." (ibid., p. 6).

[30]Ibid., pp. 14–15.

[31]Aristotle *Metaphysica* 4 in *The Works of Aristotle,* translated into English under the editorship of J. A. Smith and W. D. Ross, 12 vols. (Oxford: At the Clarendon Press, 1908–52), vol. 8. The pertinent discussion is in chaps. 4 and 5 (1005^b35–1011^a1). For a discussion of this passage see, Harold Cherniss, *Aristotle's Criticism of Presocratic Philosophy* (Baltimore: Johns Hopkins Press, 1935), pp. 77–87.

the concept of non-Being and their consequent denial of the law of contradiction. One source of their error, argues Aristotle, is their failure to distinguish knowledge and sense-experience, with the result that every sense-appearance must be true for them. Opinion is thus confounded with knowledge. This is the background to Aristotle's rather curt dismissal of Protagoras in book 10 of the *Metaphysics,* which Libavius uses in his opening assault on the Paracelsians.[32] Aristotle here asserts that what Protagoras really meant by his statement was that knowledge and perception are the measure of things, since only the man who knows or the man who perceives measures objects. This is dismissed as a commonplace, with the proviso, however, that it is understood that knowledge and perception only measure by being measured against objective reality. This reference to the *Metaphysics* leads Libavius into a brief presentation of Aristotle's faculty psychology and an attack on the subjectivity of Paracelsian teaching.

Knowledge, or the faculty of knowing, Libavius argues, involves two steps—the first sensual, the other intellectual. In the first we perceive the particular manifestations of things through our senses; in the latter we proceed to a knowledge of universals via the ideas and phantasmata stored in the imagination from past sense experiences. If our senses err in their perceptions, then it necessarily follows that our intellect will err in its grasp of reality, and the resultant knowledge will not be truth but opinion. If the Paracelsians set up their own individual intellects and senses as the measure of their judgments and pronouncements about the nature of things in opposition to the conclusions of other learned natural philosophers, both living and dead, then it behooves them to draw attention to the errors that surround their senses and to prove that they are nevertheless infallible.[33] Here Libavius abandons philsophical argument and resorts to theology. Such infallibility of sense and intellect cannot be conceded to Paracelsus; only Adam enjoyed this power when in the Garden he was allowed to name all the creatures of creation according to his knowledge and perception of them. Then, as Libavius adds, Adam's powers were intact, and he was guided by divine protection. As the Paracelsians cannot legitimately

[32] Aristotle *Metaphysica* 10. 1. 1053ª35.

[33] "Duplex per naturam homine inest notitia, seu noscendi facultas, una *sensualis,* altera *intellectualis:* Illâ noscimus singularia, quae sunt in materia haec: universalia à materia abstracta, ut ideas & phantasmata à sensibus accepta. Sensibus errantibus, necesse est & intellectum errare, fierique non *veritatem* sed *opinionem.* Quod si putant Paracelsistae suum intellectum & sensum esse mensuras, ita ut res tales & sint & iudicandae appellendaeque sint, uti ipsorum sensibus, & intellectu occurrunt, & non uti eae sunt iudicatae à Doctoribus, vel interpretibus naturae, tum vivis, tum mortuis, cavere nobis debent de erroribus circa suos sensus & probare quod absolute sint infallibiles" (*Examen Philosophiae Novae,* p. 15).

claim to emulate Adam before the Fall in their powers of intellect and sense perception, to follow them must inevitably lead down the path to the sects of Pyrrhonists, Sceptics, Heraclitians, and followers of the New Academy, all of whom deny that anything can be comprehended.[34] This is the fate that has in fact befallen the Paracelsians themselves, who have reduced all knowledge to uncertainty in their search for correspondences in the great and little world. Indeed no science is possible by this means because true knowledge resides in an understanding of the specific, inherent, and immovable causes of things, which causes are comprehended by scientific definitions and principles arrived at by reason and experience and confirmed by the judgment and experience of scholars.[35]

Libavius's other attempt to associate the Paracelsians with the pre-Socratics stems from Aristotle's criticisms of the doctrines of his predecessors in book 1 of *De Anima.* At one point in his argument in this book, Aristotle considers the doctrines of those who maintain that the soul is composed of the principles or elements of nature.[36] This hypothesis, argues Aristotle, arises out of the principle that like is known by like; hence the knowledge and perception of nature by the soul is only possible if there is identity of composition. The chief exemplar of this doctrine for Aristotle is Empedocles, although he states that Plato espoused the same view in the *Timaeus.* Libavius seizes on this passage of Aristotle and identifies the Empedoclean and Platonic doctrine with the Paracelsian macrocosm-microcosm theory.[37] For good measure he brings in Hermes Trismegistus, since in one

[34]"Adamo quidem Deus hoc concessit, ut nomina rebus imponeret secundum suam scientiam, & sensum quod tunc esset integris viribus, & regeritur praesidio divino: At Paracelso hoc tribuere non possumus, quem si sequimur facilius in Pyrrhoniorum, Scepticorū, Heracliti, Platonis, & similium factiones incidemus, qui sic mensuram rerum hominem esse voluerunt, ut *verum esset, quod cuiq[ue] videretur,* cumque hinc sequerentur *tot scientiae,* quot *capita,* omnia tandem in incertum abierunt" (ibid.).

[35]"Non enim scire est cognoscerem ex hoc illo externo comparabili aliquo modo, sed ex causis insitis, propriis, & immotis, quae causae definitionibus & principiis scientificis, ratione & experientia spectatis, iudicioque & sensu sapientum, quibus Spiritus sanctus peculiare id donum est largitum, (neque enim omnibus idem concessum est, nec omnes possunt de rebus in se, vel extra se recte iudicare) comprobatis comprehenduntur" (ibid.).

[36]Aristotle *De Anima* 1. 2. 404^b6–26, trans. I. A. Smith, in *The Works of Aristotle,* vol. 3, ed. W. D. Ross. For a commentary on this passage, see Cherniss, *Aristotle's Criticisms of Presocratic Philosophy,* pp. 293 ff.

[37]*Examen Philosophiae Novae,* p. 16. In this same passage, Libavius seizes on Croll's association of the Paracelsian *astra* with Anaxagoras's doctrine of the *panspermia.* In refutation he cites two passages from *De Anima,* i.e., 2. 4. 415^a29 and 2.5. 430^a10–15. Aristotle does not refute Anaxagoras specifically in either of these passages: the first states that all animal and plant species reproduce their own

passage of the Hermetic corpus he finds explicitly stated the same doctrine that like is known by like.[38] Thus the Hermetic-Paracelsian doctrines are linked to a pre-Socratic tradition refuted by Aristotle.[39] Such arguments abound in Libavius's polemic: particular statements of Croll are refuted by citations to the corpus of Aristotle, and where all else fails, the Scriptures are at hand.

But to judge Libavius simply on the grounds of his philosophical acumen is to miss the major point of his argument. He is not attacking Croll to defend Aristotle: he is attacking religious and intellectual enthusiasm to save his profession of schoolmastering, which he regards as the guardian of pious learning and of a stable intellectual and social order. This is the theme to which he returns again and again in the polemic against Croll. The Paracelsians wish to overthrow this order. They hold booklearning in contempt and assail the academic establishment; in so doing, they challenge the divinely instituted forms which society possesses for the transmission of knowledge. Before the Fall, God taught man directly about divine and natural things; after the Fall, such direct divine illumination became the exception rather than the rule, and God instituted in its place the normal mode whereby children learn from their parents, youths from their elders, and the unlettered from the learned. All the histories of the world, writes Libavius, bear testimony to this fact. To ignore this lesson of history is to invite intellectual chaos. In an epigrammatic cry from the heart, Libavius writes: "Take away schoolmasters and books, and let everyone philosophize for himself without the aid of either and you will have philosophical war."[40]

The keys to a sound learning for Libavius do not lie so much in the adherence to any particular philosophical doctrine; they lie in the means of assessing the teachings of the past and the present and of transmitting them to future generations. The problems are more pedagogical than philosophical. The first essential is an understanding of

---

kind (hence the notion that every object contains the seeds of every other object within itself is implicitly rejected); the second relates to Aristotle's distinction between the material cause and the formal cause which resides in the soul.

38See *Hermetica: The Ancient Greek and Latin Writings which contain Religious or Philosophical Teachings ascribed to Hermes Trismegistus*, ed. with trans. and notes by Walter Scott, 2 vols. (Oxford: At the Clarendon Press, 1924), 1: 221.

39Libavius makes the further point that if Croll did indeed learn these doctrines from Empedocles and Hermes, he learned them from men and not from the macrocosm through the light of nature.

40"Tolle magistros & libros, & sine unumquemque per se philosophari, bellam habebis Philosophiā" (*Examen Philosophiae Novae*, p. 14).

language so that the writings of scholars may be understood and interpreted correctly. This is the function of grammar. Then the doctrines of scholars must be perceived and judged correctly, which is the function of dialectic, which teaches the rules of invention and judgment. And finally the truth must be presented and transmitted in an effective form, the task of rhetoric. Grammar, dialectic, and rhetoric are for Libavius the fundamentals of sound scholarship; the skills of the word ensure the continuity of learning in acceptable social and institutional forms. To reject the normal mode of learning and to rely on inspiration is to tempt God and to leave oneself open to the seductions of the devil. At least three times in his treatise against Croll Libavius makes this point.[41] It goes to the heart of his critique of the Paracelsians. The imagination–the key faculty in the Paracelsian epistemology–when unfettered by the discipline of the word, is the haunt of Satan. It is the activity of the devil–convincing individuals that they have been divinely inspired–that has given rise to so many sects in religion and philosophy. It is the function of the schoolmaster to arrest this development.[42]

So beholden is Libavius to the word that he can not refrain from continually pointing out the Paracelsians' dependency on the verbalization of knowledge in spite of their enthusiastic rhetoric. Their ideas, too, are wholly dependent on the verbal skills of communication for their transmission. Their magic is derived from the books of Agrippa, Trithemius, Techellus, Cardan, and the *Arbatel.*[43] They commend the writings of Hermes and Hippocrates.[44] They glory in their

[41]Ibid., pp. 14, *Sectio III;* and 16, *Sectio V* and *Sectio VI.* The last of these passages reads: "Qui Deum tentant ordinario modo [i.e., discendi] reiecto, incidunt in magicas diaboli insidias, aut suis opinionibus seducuntur."

[42]"Accedit quod iuniores ob imperitiam, & assentiendi, tum levitatem, tum temeritatem, à sophistis facile seducantur, quodque Satanas simulans angelum lucis & apostolos suos passim distribuens, mendaciis veritatē opprimere satagat, quod manifestum est tot sectis sacrorum & profanorum, quin raticiniis antiquorum, & experientia ocularique intuitione, adeo ut quo provectior aetate est mundus, eo certaminum sit maior copia" (ibid., p. 14).

[43]Ibid. Techellus was a legendary Jewish sage and magician who was allegedly the source of ancient knowledge about the magic of stones and carved images (See W. Pagel, "Paracelsus: Traditionalism and Medieval Sources," in *Medicine, Science and Culture: Historical Essays in Honor of Owsei Temkin,* ed. Lloyd G. Stevenson and Robert P. Multhauf [Baltimore: Johns Hopkins Press, 1968], pp. 65–66; and idem, "Paracelsus and Techellus the Jew," *Bull. Hist. Med.* 34 [1960]: 274–77). Jerome Cardan (1501–76) was the famed Italian mathematician and voluminous author on magic (see J. R. Partington, *A History of Chemistry,* 2: 9–15). The *Arbatel* was an anonymous digest of magic which first appeared in Basel in 1575 (Thorndike, *A History of Magic and Experimental Science,* 6: 457–58).

[44]*Examen Philosophiae Nova,* p. 15.

conversations with artisans, which is to learn from men.[45] Croll himself is a disciple of Paracelsus and Severinus.[46] Libavius even taunts them for seeking out academic degrees, although he cannot resist the aside that these can be bought and do not necessarily testify to the absence of stupidity.[47]

The drift of Libavius's polemic so far is clear. In their claims to special revelation in the lights of grace and of nature, the Paracelsians were playing with fire—hellfire. Satan was ever ready to pose as the angel of light and seduce the unwary, particularly the young. The prevalence of such enthusiastic notions was threatening the very fabric of society and its institutions. Here Libavius echoes sentiments he expressed in his first chemical epistle to his young friend in 1595. The gathering storm of confessional strife (he was writing only three years before the outbreak of the Thirty Years' War) only gave fresh urgency to his appeals. His remedy was a humanist one: a renewed respect for the fundamental institutions of society—the family, the school, the university, the established Church—which embraced the collective wisdom of the ages. He attempted valiantly to engage the Paracelsians and enthusiasts on these grounds, by attempting to locate and refute their arguments in the light of institutionalized knowledge. It is this which gives particular poignancy to the opening of his ninth section in the polemic against Croll: *at operam fortasse ludimus* ("but perhaps we are only playing at games").[48] He realized here that his efforts were futile: the Paracelsians could not be so engaged, for they did not speak the same language. What was referred to as astronomy was not what Ptolemy and Copernicus had taught. Words did not have the same meaning for them as for others; hence even dialogue was impossible. Here Libavius brings himself and us back to the fundamental issue—the meaning and signification of the Word.

Croll's chemical philosophy was predicated on the search for the Word in nature—the divine Word. This Word, which stemmed from

45"Cum etiam gloriantur de suis peregrinationibus, & colloquiis cum variis artificibus, an hoc non est discere ab hominibus?" (ibid., p. 16).

46Ibid.

47"Non aliter discendum esse ipsi Paracelsistae suo exemplo docent. Quis enim ex illis non didicit literas ab aliquo praeceptore seu publico, seu privato? Quotus quisque est qui sibi non venetur titulos utriusque docturae ex privilegis Academicis? Quos enim bullatos vocant, ii nomine sunt non re, & lege Doctores, saepeque fiunt tales pro nummis, quibus non tollitur asinus" (ibid.). The reference to *titulos utriusque docturae* in this passage is possibly directed at Paracelsus's frequent description of himself as *Doctor in utraque Medicinae*, a degree awarded exclusively in his lifetime in Northern Italy. See Pagel, *Paracelsus: An Introduction to Philosophical Medicine in the Era of the Renaissance* (Basel and New York: S. Karger, 1958), p. 10, n. 26. The pun on *bulla* is typical of Libavius's wit.

48*Examen Philosophiae Novae*, pp. 17–18.

the single creative utterance of God the Father, was made reaccessible to man after his fall, by the reincarnation and redemption of the Word made Flesh in Christ. Croll's aim was to direct attention to the primeval text of things contained in the Book of Nature. The book was not a text of words about things but a book of signs which revealed the hidden nature of things and bore testimony to their power. This power resided in the similitudes, analogies, and sympathies of the things themselves, all reflected through man, who, as the microcosm, was the focal point of all the analogies in the great world. These similitudes and analogies were themselves revealed through the signatures of things, the ciphers which bore the mark of the hidden resemblances and made possible the reading of the text. The signatures were themselves similitudes, of the same nature and essence as the resemblances they revealed, so that it was impossible to decipher the text without calling into play the powers of sympathy which the text manifested. Similitude and analogy were not for Croll figures of speech which illuminated the essentially incomparable; they were the very fabric and glue of the universe and the means by which it spoke. Croll's text, in turn, was not a discourse about the nature of things, but a key by which nature spoke to us with the echoing power of the divine utterance of creation and redemption which remained ineffable for man. Croll's text ineluctably invited his reader to step out from his dead book into the living world of experience.

The vaunted muteness of the Paracelsians in the presence of the ever echoing voice of nature placed them, in Libavius's view, beyond the pale of rational and reasonable human beings. They were to him madmen (*vesanus* and *delirus* are adjectives he frequently employs to describe Paracelsus and Croll) locked up in the phantasies of their imagination. All knowledge begins for him with the articulation of words—the words of men—carried through space by speech and transmitted through time by writing. For Libavius language was of human construction and was governed by man-made rules of grammar, dialectic, and rhetoric. It was the instrument by which man had civilized himself in this world and through which he learned of his salvation and spiritual destiny in the words of the Scriptures. For Libavius no relic of a primeval text of nature existed, not even in the Hebrew Cabala. Indeed such a text never existed. Adam imposed (*imponeret*) names upon the creatures of creation according to his knowledge (*secumdum suam scientiam*).[49] He did not simply read the transparent text of nature, but he himself gave its particulars their denominations. Likewise fallen man must strive within the limits of his impaired faculties to give

[49]See n. 34, above.

names to things. Nature stands mute awaiting man to call it into consciousness through the power of his language to discriminate it. Significantly, Libavius does not mention the episode of the tower of Babel; his historical account of language does not entertain the recovery or reconstruction of the universal language of divine origin of the pre-Babel era. There was no second fall—no fall of language; at least it does not figure prominently in Libavius' thought. He seems quite content in the univocal culture of the Latin West.[50] Adam's fall was one of the senses and of the intellect, which impaired his power to denominate things and therefore to comprehend them. The history of mankind since that initial catastrophe had been the epic of man's striving to grasp the meaning and destiny of his existence through the refinement of his words.

This man-made language of which Libavius speaks operates in a wholly different way from the language of nature which Croll evokes. It defines, divides, distinguishes, and establishes criteria for judgment—a judgment which separates things. It seeks to discriminate knowledge, whereas Croll's language sought to reveal and express the affinities and resemblances of things.[51] Language for Libavius is a knife, ever sharpened by use, which dissects knowledge into discrete segments in the dimensions of space and time.

This language operated in a wholly different way from the magical language of similitudes. It is a realization of this which suddenly surfaces in the ninth section of Libavius's critique of Croll and which cuts through all of his philosophical arguments. "The Cabala," says Libavius of Croll's archetypical language of signs, "is a falsehood and a deceit. For it presents things, not as they are, but as they are compared with other things in an indeterminately external fashion. Thus we are not able to know what constitutes a thing, for the gateways [to knowledge] are surrounded by deceiving images."[52] The continuing search of the Paracelsians for the similitudes of things obfuscated the path of knowledge and made exact sciences an impossibility. True knowledge is the product of man's effort to develop and use the discriminatory power of the word, and not the result of an

[50]The cacophony of tongues which was visited on man as a retribution for his presumption in attempting to build the tower of Babel was seen in medieval times as God's punishment for a second sin of pride. The gift of tongues given to the apostles was regarded as the redemption of this second fall.

[51]See the quotation in n. 35, above.

[52]"*Cabala ista mendacium est & vanitas.* Non enim res offert uti sunt, sed uti comparantur cum aliis qualicunque modo externo. Ita non possumus scire, quid in re sit: foris enim circa imagines fallaces occupantur" (*Examen Philosophiae Novae,* p. 18).

accumulating gloss of interpretation on the text of nature replete with similitude and metaphor. In Libavius's scheme of the world, conceived in terms of word skills, magic and the Cabala were associated with the figures and tropes of rhetoric and poetry. The world and language of similitudes were properly the province of the poet; the conceits of imagery were banished to the realms of the literary imagination.[53]

The primary target of Libavius's attack was the open-ended discourse of similitudes inherent in the Crollian world view, and in particular the "divine" analogy of the macrocosm-microcosm, which was its epistemological basis. However, he did not expunge analogy completely from philosophical discourse; rather he restricted it to a well defined and technical form of argumentation which had its roots in medieval scholasticism.[54] In this context analogy was a form of probabilistic argument which invoked an appropriate juxtaposition of natures or ratios (*rationum convenientia*). This was the analogy of proper proportionality, which distributed an intrinsic property or attribute shared between two or more terms in a proposition to illuminate the degree of participation of each term. Libavius's rather crude example of numerical proportions demonstrates this. He argues that the analogous ratios of the numbers *4* and *2*; and *12* and *6* in no way imply that the entities which may be represented by these numbers are the same, nor that the equivalent terms in each ratio participate to the same degree in the shared attribute (in this case number).[55] The analogy of proper proportionality was not restricted of course to quantifiable categories, but even in the very limited example of Libavius, the proper function of such analogy can be detected. This was to emphasize the *difference* in the degree to which the related terms participated in a common attribute. Analogy of this type served to differentiate elements of a proposition and was not intended to reveal hitherto unsuspected identities or relationships. Indeed the very possibility of forming such an

53"Praetera ista Magia & Caballa nihil est aliud, quam Rhetorica & Poetica tropologia, nata ex comparatis logicis per qualescunque similitudines, quo modo possum Deum in hominem transformare & hominem in Deum permutareque bestialia rationalibus, & alia mirifica plus quam Ovidiana metamorphosi producere" (ibid.).

54Analogical argument in medieval scholasticism is an important topic with a vast secondary literature. For an excellent introduction see Marcia L. Colish, *The Mirror of Language: A Study in the Medieval Theory of Knowledge* (New Haven and London: Yale University Press, 1968), pp. 209–23, and the extensive bibliography therein.

55"Quid enim est *analogia* quam rationum convenientia? Rationes sunt in rebus & cognosci debent per se prius, quam convenientia statuat. Non autem omnis analogia par est, aut similis, multo minus analogia idem concludit. Non enim si proportio eadem est inter 2. & 4. 6. & 12. res ipsae sunt eaedem aut numeri. Hic est pudor Paracelsicus" (*Examen Philosophiae Novae*, p. 21).

analogy depended upon an a priori knowledge of the appropriately shared attribute or characteristic, which the analogy sought to define more precisely. From this standpoint, analogy could not possibly enjoy primary epistemological status, which is Libavius's main point. In scholastic philosophy, analogy of proper proportionality most commonly served to differentiate the common attributes of God and his creation, such as being. Analogy employed in such instances was intended to manifest the inherent distinctions in such comparison and at the same time to provide an imprecise intellectual grasp of a relationship which transcended the powers of reason alone. Libavius continues to recognize the necessity of this use of analogy in ascending from a knowledge of ourselves and nature to a knowledge of God, although he cautions that such analogies can be ambiguous and misleading—an expression, no doubt, of his Lutheran suspicion of an overconfident natural theology.[56]

The relationship of man's sensory experience of nature and man's accumulated verbal knowledge is one of the more elusive elements of Libavius's thought, but once seized, it provides the key to his epistemological position. It is the context rather than the content of his statements which gives the clue. Like Croll, Libavius refers to the Book of Nature, but this occurs in two places where he is establishing pedagogical norms and responsibilities. In the first instance he states that once the rudiments of learning have been laid down through teaching (the rudiments of language, naturally), the contemplation of nature ought not to be neglected, for by this means the student can gain knowledge (*notitia*) and examples (*documenta*) of the most abstruse and worthy things, which confirm what our elders have bequeathed to us from their reason and experience.[57] In other words, the Book of Nature is a supplementary text which enhances the student's confidence in the printed text used in the classroom. The second instance again occurs in the passage where Libavius defends the traditional pedagogical mode of learning as the sole method of gaining true knowledge. Following the familiar invocations to learn letters and the

[56]"Hoc lumine [i.e., lumen naturae] utimur naturaliter in contemplatione nostrûm, & aliarum creaturarum ad DEVM inde cognoscendum. Id sit & per essentiam, & per similitudines. Illa notitia si haberi integra possit, certior est: haec spuria & ambigua: utraque tamen in hac vita *imperfecta* . . . " (ibid.).

[57]The Book of Nature is here coupled with the Book of Scriptures: "Interim percepti per doctrinam initiis non negligenda est naturae contemplatio. Nam & in hac sunt rerum abstrusarum, & notitia dignissimarum documenta, quibus etiam ex parte firmatur id, quod maiores nostri nobis reliquerunt ratione & experientia constitutum. Itaque & *naturae liber nobis est & scripturae.* Utrobique vero iuniores indigent informatione maiorum sapientia & peritia pollentium" (ibid., p. 14).

true dialectic of invention and judgment, he urges a comparison of the teachings of scholars measured against the Book of Nature, which never changes.[58] The attention moves from the texts of men to the Book of Nature and back again to what man has written about nature in his textbooks. Again the Book of Nature is secondary to the printed word, a visual aid supplementing the textbook.

Quite clearly what has happened here is that pedagogical practice has completely swallowed epistemological theory. The operations of the intellect have been identified with the activities of school; in short, the mind has been confounded with the classroom. Libavius all along equates the acquisition of knowledge with the process of learning, his professional instincts and commitment winning out over his philosophical acumen and allegiance. The brief obeisances to Aristotle's faculty psychology are only loyal gestures made in the midst of the more pressing concerns of dunning knowledge into the heads of schoolboys. The process of learning and hence of gaining knowledge is bounded by the walls of the classroom. The initial stages are devoted to the mastery of the skills of the word—speaking, reading, and writing. These are followed by the syntax of language and argumentation in grammar and dialectic. And finally the substance of accumulated learning is approached through its rhetorical presentation in the texts of past ages and, increasingly in Libavius's generation, through the systematic presentation of the subject matter of the curriculum in the textbook. In this last phase the teacher's responsibility is to relate the substance of the text to experience outside the classroom in justification of the contents of the textbook. Experience has significance only in its relationship to the spoken and printed word. This sequence not only determines Libavius's theory of knowledge; it also colors his interpretation of the historical development of philosophy. Having defined the twin faculties of sense and intellect as the light of nature, he goes on to state that the "ancients" first gave this name to dialectic, which they regarded as embracing the whole function of reasoning; to this conception of the light of nature the Peripatetics added sense-experience.[59] Even historical evolution follows pedagogical practice. Knowledge radiates out from men in the form of their words to embrace experience. The overwhelming pedagogical bias of Libavius's viewpoint is manifest in the way in which he slips easily from the words *dicere* and *docere* to

[58]"Disce Dialecticam synceram, quae est & inveniendi & iudicandi regula. A Deo postula Sapientiam, & luminis naturae illuminationem. Ausculta sapientes eorumq; placita confer cum placitis aliorum, & rebus ipsis. Habes librum naturae perpetuo eiusdē, & sibi per se conformis" (ibid., p. 16).

[59]"Veteres *Logicam* seu *Dialecticam* omne rationis officium complexam lumen naturae vocarunt. Ei Peripatetici iunxerunt sensus" (ibid., p. 21).

discere, and equates all three with *scire.* Only that which can be said (and written) can be taught; only that which can be taught can be learned; and only that which can be learned by being taught can be known.

Libavius is in fact a product of and a spokesman for what Walter Ong, in his book on Ramus, has admirably described as the "pedagogical juggernaut" of the sixteenth century.[60] Our schoolmaster's place in this phenomenon will be discussed in the next chapter, but a few anticipatory remarks are appropriate here. Libavius belonged to the golden age of schoolmasters, that tribe who swarmed through European society in the second half of the sixteenth century to capture, form, and tyrannize the minds of *burgelische* youths. In Protestant societies they were spawned by the example of Melanchthon and Ramus. They derived their authority from state-established religion whose buttress they were. Their *fora* were the public schools, and their tools were their "methodized" textbooks which reduced all knowledge to a form suitable for ready digestion by the delicate tracts of immature minds. It is this last which is of greatest significance in Western intellectual history. Their enterprise exploited and indeed was made possible by the availability and diffusion of the printed book. All that was knowable was reduced to the systematic order of their textbooks, and minds were trained to identify knowledge with the layout of words on the printed page. The system was designed to focus eyes on the text and to facilitate the immediate transference of the visual components on the page to the intellect with the minimum of interference and delay. A significant sector of European society came to see the world as a series of textbooks. The oral and aural components of learning were reduced to viva voce exercises designed to imprint the image of the page on the memory.

The principal casualty of the system was the imagination. It was a dangerous faculty, which interfered with the efficient process of learning in that it impeded that swift passage of knowledge from the page to the mind. In one sense the function of the imagination was usurped by the schoolmaster, who came to moderate the transference of knowledge from sense to intellect by riveting the attention of his pupils to their textbooks. In another sense, the imaginative faculty was rendered redundant in the process of learning. There was no longer any need for the elaborate apparatus of image-constructions, which formed the art of memory, an apparatus essential to the whole academic

[60]Walter J. Ong, S.J., *Ramus: Method and the Decay of Dialogue* (Cambridge, Mass.: Harvard University Press, 1958), esp. chap. 7, "The Pedagogical Juggernaut," pp. 149–67.

process before the widespread diffusion of the printed text.[61] It is perhaps no accident that the principal vehicle for memory training became, in Libavius's generation, the learning by rote of rhetorical speeches and poetry. Here was the proper place for images and similitudes.

Hence when Libavius speaks of the Book of Nature he has in mind the clear-cut, well ordered textbook of the classroom. Croll's book, on the other hand, was at once more mysterious and more darkly sensual. In the first place, it had not yet been bound; pages remained to be opened to reveal their hidden mysteries. Neither were its visible ciphers meant to fly immediately to the intellect; they were intended to float in the imagination, where they would gradually reveal their dense layers of meaning. Nor was the book limited by the visual sense. The power of the words of nature was ultimately derived from the spoken Word of God. Paracelsus urged his followers to "overhear" (*ablauschen*) nature.[62]

To appreciate the full force of Croll's metaphor, it is helpful to recall its medieval roots. As Curtius reminds us, the metaphor of the Book of Nature was a commonplace of medieval preaching.[63] It was intended for a largerly illiterate audience with no access to the book of words. For them the book was a repository of mysterious symbols whose meaning had to be interpreted for them by means of the spoken word, the only form of the word they fully understood. It was a metaphor whose significance can only be appreciated in the context of a semiliterate society. It is in this spirit that Paracelsus and Croll urge their followers to abandon the words of men in academic texts, embrace the mentality of the illiterate peasant and artisan, and search out the divine hieroglyphs of nature with the appropriate sense of awe and mystery.

But the different doctrines of the Word which Croll and Libavius espouse do more than define opposing epistemological positions: they reach down and inform radically contrasting views of man, his history, his society, and his relationships with nature and with God. For Croll, the Word coursed vertically through the cosmos and linked men actively and individually with God through nature. It wrested man from his earthbound existence and elevated him to a level from which

[61]Frances A. Yates, *The Art of Memory* (Harmondsworth: Penguin Books, 1969).

[62]Pagel, *Paracelsus*, p. 60.

[63]Ernst R. Curtius, *European Literature and the Latin Middle Ages*, trans. Willard R. Trask (New York: Pantheon Books, Bollingen Series 36, 1953), pp. 319–26.

he could conduct an ongoing dialogue with the divine—an ineffable dialogue, to be sure, but one that was manifest in works. Men were not thereby freed from social responsibilities, but their personal witness to the Word was thought to transcend obligations to the institutional forms of society and religion. Such a view was explicitly critical and implicitly destructive of the established social and religious order, hence its close association through the sixteenth and seventeenth centuries with both radical social criticism and irenical religious sentiment. It derived its legitimation from personal divine illumination.

As a world view it was inherently revolutionary, containing within itself the seeds of continual renewal. It was a renewal, however, that did not wait upon the actions and efforts of men but rather was ensured by the continuing revelation of the divine plan in the Book of Nature.[64] Nature was not a complexity of entities with their own inherent and interdependent cycles of becoming and passing away which in their constancy and regularity provided fixed parameters by which man located himself in space and time. Rather, nature was the vehicle of divine messages, which revealed knowledge appropriate to every age. It was in part on this basis that the Paracelsians rejected so confidently in their rhetoric, if not in their practice, the doctrines of past ages—not only were these for the most part founded on erroneous principles, but any practical value they may accidently have contained had ceased to be relevant. One of the most constant justifications for the innovative Paracelsian therapy was that these medicaments were the novel and appropriate response to the "new" diseases which had been revealed by nature in contemporary European society.[65] The elect had to be ever attentive to the renewal of the message of the Word in the Book of Nature and to be prepared to bear witness in their own self-renewal and in their works.

Libavius, on the other hand, is a stout defender of an established social order which enshrined what today would be characterized as middle-class virtues and values. His was a world that celebrated the family, the school, the university, the state, and established religion, all of which institutions were dependent on the skills of the word for their

64"Doubtlesse there are more secrets yet concealed in the Treasures of Wisdome and Nature then we perceive, which (being ordained for Times and Nations, by an immutable decree to the end of the world) are to be sought out by wise-hearted men" Croll, *Philosophy Reformed and Improved in Four Profound Tractates. The I. Discovering the Great and Deep Mysteries of Nature: By that Learned Chymist and Physitian Osw: Crollius. The Other III. Discovering the Wonderfull Mysteries of the Creation, By Paracelsus: Being His Philosophy to the Athenians. Both made English by H. Pinnell, for the increase of Learning and true Knowledge* (London, 1657), p. 127.

65Pagel, *Paracelsus,* p. 139.

functioning and their continuity. All were in their separate ways institutes of learning. Life was one giant pedagogium where deference was paid to age and experience, and elders taught their juniors. It was through this social order, held together by words, that Libavius believed man overcame the consequences of his fall, when he lost his knowledge and power of dominion over nature, a power symbolized by his ability to name the creatures of God's creation. Adam, the individual had fallen. Mankind collectively would rescue the situation, in part by its ability to deploy and refine its skills of language to organize experience and establish social forms and institutions, which ensured the orderly transmission of these skills. Language was the vector of material prosperity, cultural progress, and religious fulfillment. We find Libavius in one of his chemical epistles quite naturally and spontaneously equating the origins of man's arts and his political organization with the development of language.[66] Man's words were the very fabric of his society, linking men across space and through time. Language stretched horizontally through the cosmic order, binding men together but setting them apart from the rest of creation. Even God spoke to man through man's words in the Scriptures. The Word did not resound vertically through the cosmic order, summoning man directly to God through nature. Between God and man stood nature, not as a harbinger of God's revelation, but as the mute and constant testimony of God's beneficence, awaiting man's endeavor to rename it and thereby to regain his dominion over it. It was this constancy of nature which ensured the progress of knowledge, for it enabled a continuity of dialogue to be maintained through historical time with reference to a common experience.[67] This dialogue would not be free from contention, but Libavius, ever the polemicist, argues that such contention is a

[66]Libavius, *Rerum Chymicarum Epistolica Forma ad Philosophos et Medicos quosdam in Germania excellentes descriptarum Liber Primus, in quo tum rerum quarundam naturalium continentur explicationes ingeniosae; tum Chymiae disciplina pyronomica, sceuastica & vocabularia cum quibusdam inter arcana habitis declarantur fideliter. Autore Andrea Libavio Med. D. Poeta & Physico Rotemburgo tuberano* (Frankfort, 1595), 1: 107–8. In the polemic against Croll Libavius asserts that Prometheus did not snatch the knowledge of man's arts from the sun but learned them in the workshop of Minerva and Vulcan (i.e., he was *taught* them): "Ideo & Prometheus fingitur eas [i.e., artes] ex officina Minervae & Vulcani transtulisse, non ex sole rapuisse" (*Examen Philosophiae Novae,* p. 21). He did not go as far as Ramus, however, in identifying Prometheus's "fire" with dialectical principles (see Ong, *Ramus,* p. 172).

[67]See the last sentence of the quotation in n. 58, above. Libavius devotes much of *Sectio V* of the polemic against Croll to a defense of the thesis that nature has not changed through historical time and to a rejection of the Paracelsian idea that their new medicaments were justified by new diseases. "Non est tanti Gallia lues, morbus Ungaricus & similes, ut oporteat medicinae principia immutare.

necessary dynamic of knowledge.[68] His principal frustration with the Paracelsians was that he could not engage them in such dialogue—they did not speak the same language. Furthermore, it was the rupture of this vertical link between man, nature, and God which made man's exploitation of nature possible. Man was not attendant on the divine message of nature; he had God's sanction in the Scriptures to develop nature to his own useful ends. In order that this be accomplished in socially constructive ways, it was necessary that man's arts be reduced to systematic presentation in words. Libavius's inevitable response to the Paracelsian enthusiasts' exploitation of the chemical arts had been to write a textbook of chemistry.

Fuerunt olim isti morbi, et iam sanantur iuxta praecepta medicinae priscae, quae nova remediorum praeparatione minime abolentur, ut taceamus chymia repeti à multa vetustate, neque nova esse ad novos morbos remedia" (*Examen Philosophia Novae*, p. 16).

68"Turpe est stationem suam deserere ingruente hoste. Turpe ergo & medicam, physicamque veritatem à suis professoribus destitui sophistica gliscente. Dissentiunt inter se Doctores & libri: at hic dissensus ex collatione utilis est ad lumen veri agnoscendum" (ibid., p. 15).

CHAPTER VI

CHEMISTRY INVENTED

Andreas Libavius was a Ciceronian through and through. In one place in his chemical epistles he refers to Cicero as "that wise philosopher and orator of the Romans" and compares Plato unfavorably with him.[1] The conjunction of philosopher and orator is significant. The celebration of the orator as philosopher, and of Cicero in particular, stems from the Lutheran humanist tradition of Melanchthon. For Melanchthon the arts of speech were fundamental—everything was subordinate to rhetoric.[2] The import and justification of all knowledge was that it be socially useful and morally uplifting in matters both secular and divine; this was only feasible if wisdom were matched by eloquence. Thus the theologian was preeminently the preacher who expounded the Scriptures to his congregation; the statesman by his eloquence formed public opinion and instituted good laws; and the lawyer exercised his forensic skills in seeing that these laws were executed fairly. This humanist ideal of society envisioned a Christian commonwealth which functioned primarily on the skills of the orator. But the keystone to the whole structure was the schoolmaster, the orator *primus inter pares,* who inculcated the rudiments of letters and trained his pupils in the arts of words in order to equip them for their proper station in life. While it is true that in the sixteenth-century realization of this ideal the printed word increasingly came to challenge, and eventually to overwhelm, the spoken word, the forms and organization of discourse even in writing continued to be strongly influenced by the syntax of the oratorical art.

[1]Libavius, *Rerum Chymicarum Epistolica Forma ad Philosophos et Medicos quosdam in Germania excellentes descriptarum Liber Primus, in quo tum rerum quarundam naturalium continentur explicationes ingeniosae; tum Chymiae disciplina pyronomica, sceuastica & vocabularia cum quibusdam inter arcana habitis declarantur fideliter. Autore Andrea Libavio Med. D. Poeta & Physico Rotemburgo* (Frankfort, 1595), hereinafter to be cited as *Rerum Chymicarum Epistolica,* 1: 56.

[2]See Q. Breen, "The Subordination of Philosophy to Rhetoric in Melanchthon," *Archiv für Reformationsgeschichte* 43 (1952): 13–27.

It was this essentially Ciceronian culture which Libavius lived and breathed. This is manifest throughout his works, but most especially in his chemical epistles, which were designed to absorb the chemical art into this cultural ideal. One would expect to find a refined Latinity and a fair sprinkling of literary classical allusion in the writings of a professor of classical history and poetry. But Libavius's epistles are more than a vehicle for erudition: they are informed throughout by quite explicit humanistic ideals. In the first place, the very form of a collection of open epistles is Ciceronian. He writes publicly to his former pupils, friends, and academic colleagues inviting their response to his thoughts on how chemistry can be made respectable and brought together in a form suitable for teaching.[3] In the dedication of these epistles to the senators and consuls of Ratisbon, a self-governing commonwealth within the empire, he reveals his ideal of imperial polity—a constellation of city-states bound to the authority of the emperor in which the *monarchia* cooperates with the *aristocratia* for the best form of political government.[4] (Libavius always refers to the state by that sterling Circeronian term, the *res publica*.) This dedicatory choice was deliberate and was intended to contrast with the insinuation of the Paracelsians into the princely courts of Germany. Libavius had a deep suspicion and contempt of courts: they were the breeding grounds of secretiveness and intrigue and hence of Hermetic and Paracelsian cabals. In one of his letters he invites his reader to examine the records of the courts of Germany to discover how much money was being thrown away on the likes of the Paracelsians.[5] But the influence of Ciceronian humanism runs deeper than Libavius's views on public polity. We have drawn attention in previous chapters to his basic belief in the imperfection of human knowledge in this world and attributed it to a cautious Lutheran piety. This it was in part, assuredly; we find similar attitudes in Melanchthon.[6] But in both instances this was a Christian gloss on Cicero's moderate Scepticism and Stoicism. Indeed we find this suggestive association in an epistle in which Libavius stresses the necessity of making a start in formulating the new chemistry:

> . . . no one can acquire the sort of wisdom which the Stoics have postulated: nor can anyone be an orator of the type described by Cicero since the ideals of the arts

[3] *Praefatio ad Lectorem,* in *Rerum Chymicarum Epistolica,* 1: [7–8].

[4] "Ubi enim Aristocratia cum Monarchia, laudatissimis, nimirum gubernationū politicarum speciebus ad formam optimi Imperii iucūda contemperatione conspiravit" (*Dedicatio,* in ibid., sig. 6 recto).

[5] "Evolve mundi historias, & praesertim Germanorum principum argiva; vix numerare poteris preciosissimorū thesaurorū pondera, quae à Paracelsitis nefariè sunt interversa" (ibid., vol. 1: 37).

[6] See Breen, "The Subordination of Philosophy," pp. 21–22.

are impossible to attain. But it is important to have such ideals as targets to aim at and strive for with all our strength. . . . Thus in chemistry if a final resolution of all the parts is not achieved, no one need be ashamed to have made a start and to have done something, because it will instruct us about divine power, and will bring some benefit to the public welfare.[7]

Elsewhere when he was discussing the necessary condition for progress in learning, Libavius stressed an innate talent, practice (*habitus*), and a dedicated perseverance towards one's chosen goal.[8] All of these requirements are those described by Cicero for the formation of the ideal orator alluded to in the above quotation. Incongruous as it may seem in retrospect, Libavius's ideal of the true chemist was influenced by Cicero's ideal of the true orator as set out in *De Oratore.*[9]

In his second letter Libavius anticipates some of the difficulties which his former pupil will confront if he wishes to become a true chemist. The first of these is language:

> The true chemist does not like neologisms. The Paracelsian likes nothing more. To hide in a perplexity of words like the Delphic Demon is the best that these artful dodgers are capable of. But this should not be taken to absurd extremes and [those words] which have been taken over from the Arabs and the Egyptians have been made more pliable now by frequent use. Let the latin-speaking practitioner speak more latinly when he can. He dare not change everything, however, for change of this kind obscures the subject-matter.[10]

The question of terminology was paramount to Libavius in his endeavor to formulate a genuine chemistry. He returns again and again to the problem in his letters, and one entire epistle (letter 18) is devoted to the question.[11] In this letter, Libavius offers an eloquent and witty critique of the inanities (*vesania*) and obtuseness (*stoliditas*) of the language of similitude of the Paracelsians as well as of the pictorial emblematic tradition of alchemy. As he states at the end of this letter, he will leave this sort of thing to the crows, the crow being a common alchemical emblem for any process which blackened the imperfect metal prior to transmutation. The problem is how to deal with the peculiar terminology of the literature on the chemical arts and the vocabulary of the chemical artisan. If this difficulty is not confronted, then not even a beginning can be made; but if it is mastered, then one is

[7] *Rerum Chymicarum Epistolica,* 1: 54.

[8] See for instance Libavius, *Examen Philosophiae Novae,* in *Appendix necessaria Syntagmis Arcanorum Chymicorum* (Frankfort, 1615), p. 16.

[9] Cicero *De Oratore* 1. 25–35.

[10] *Rerum Chymicarum Epistolica,* 1: 44–45.

[11] Ibid., 1: 162–71.

halfway home.[12] In his analysis of this problem Libavius shows a perceptive appreciation of the mentality of the artisan:

> Nowhere is the proper word worth more than in the arts. This has not arisen fortuitously, but stems from the experience, works and motives of trade. Thus much careful attention should be paid to this if you wish to make good progress. New words are sought [so that] outsiders cannot follow the chemical expressions. If these same outsiders enter a shoemakers or a metalworkers, do you think they will understand the workers talking amongst themselves? Will they not rather leave than provoke an argument? They [the artisans] will write down their instructions for their sons and apprentices in the way that Arabs speak. One must get used to this. However, I cannot deny that occasionally something has been deliberately obscured or twisted from its original meaning: [but] this does not arise from jealousy of the art, but from the fear of unfit men.[13]

This, in Libavius's eyes, was very different from the deliberate distortions of the Paracelsians. The artisan was after all making significant contributions to public welfare and was only seeking to protect his livelihood. In the preface to the epistles Libavius points out that some readers will note his use of "popular" words, which derives from the fact that he has found it necessary to speak with many artisans. But this should be the first lesson: namely, to learn any art one must first master its peculiar terminology.[14] Still Libavius was too thoroughgoing a Latinist to leave matters there. He refers to the problem frequently, particularly in epistle 18. Here he states that if all arts were to be decried because their language did not have Ciceronian eloquence, then even medicine, with its peculiar terminology full of Arabic terms, would be discredited. Did not Cicero himself expand his vocabulary with Greek words? And did not Plautus use whole Greek phrases and maxims, as well as Punic words, in his plays? Thus, concludes Libavius, we can allow words of Arabic, Egyptian, and Chaldean origin in our science. Such is his respect for the word that he fears translation of these non-Latin terms which have become established by long usage would lead to confusion about their meaning. It is better to give them Latin terminations just as Cicero did with his coinings from Greek.[15] In his policy of adopting the particular vocabulary of the arts in his scholarly discourse, Libavius was echoing the sentiment of Melanchthon, who earlier had written:

> In each particular art there are certain technical expressions appropriately used by masters thereof. So to philosophers and theologians their peculiar expres-

[12] Ibid., 1: 48–49.

[13] Ibid., 1: 49.

[14] *Praefatio ad Lectorem,* in ibid., 1: [9].

[15] *Rerum Chymicarum Epistolica,* 1: 163–64.

sions are to be allowed, as also to architects and painters. But at the same time the rest of their diction ought to conform to usage and to known terms, and the joining together of terms should follow custom.[16]

Their mutual brand of humanism made correct Latinity subservient to social utility.

But Libavius's ambition was to do more than simply bring an acceptable Latinity to chemical discourse. Earlier in the sixteenth century a fellow German humanist schoolmaster and physician, Georgius Agricola (1494–1555), had written extensively in Latin on all aspects of the art of mining and had developed an appropriate terminology for the vernacular expressions and technical vocabulary of the artisanal community amongst which he lived and worked for a significant period of his life.[17] Agricola's great work, the *De Re Metallica,* published by Froben at Basel in 1556, was a conscious effort to provide an exhaustive treatise on the metallurgical arts to complement such Roman encyclopedias of the arts as those of Varro, Cato the Elder, Vitruvius, and Celsus. Agricola's achievement, although magnificent, was based wholly on classical models. Libavius's project, on the other hand, was of quite a different order. He was seeking to locate the common operations and prescriptions of the chemical arts, or the arts dependent on chemical technique, and to organize them into an independent branch of natural philosophy which would produce benefits to society in its works.

In one of his letters Libavius provided some fine rhetorical flourishes to point up the originality of his project.[18] Following a long discussion on the history of man's social organization and on the origin of his arts, in which he attempts to reconcile the biblical account with those of classical authors, he muses on the strange fact that, in spite of the many subjects investigated and brought forth (*accersita*) in the ancient world, chemistry had never been set out in the form of an art, with appropriate axioms as befits a demonstrative science. This is even more surprising given the antiquity of the metallurgical arts. "What inertia has seized the centuries that this subject alone has been neglected when so much else has been seized hold of and furnished with its own cultivation?"[19] There follow some pious comments about the abstruse nature of chemistry, which has hindered its "invention" and

[16]Cited in Q. Breen, "Melanchthon's Reply to G. Pico della Mirandola," *J. Hist. Ideas* 13 (1952): 418.

[17]On Georgius Agricola, see Partington, *A History of Chemistry,* 4 vols. (London: Macmillan, 1961–70), 2: 42–55.

[18]"De Constitutione artis Chymicae," in *Rerum Chymicarum Epistolica,* 1: 107–16.

[19]*Rerum Chymicarum Epistolica,* 1: 112.

practice. Unfortunately its proponents have not set out the art in "scientific" precepts but have covered it over with mystical statements and surrounded it with a divine aura. Probably this lack of proper organization is the reason it has not yet been wholly "invented" or solicitously cultivated; for the works of the ancients include teaching (*doctrinam*) about minerals, and much time and energy has been devoted to transmutation. "Can it be," Libavius wonders, "that books have perished in wars or some other disaster?"[20] These historical reflections aside, Libavius's principal concern was with the contemporary state of writing on chemical subjects.

When Libavius turned to the accessible contemporary literature on the chemical arts, he found it either limited in its content or lacking in organization and logical method. He recognized first of all two traditions stemming from the late Middle Ages: alchemy and distilling. The many contemporary alchemical works were based upon the example of the Latin Geber, who had given many precepts about alchemy, including its definition and division. These, however, were restricted to the subject of the nature and transmutation of metals.[21] The numerous works on distilling he traced back to the influence of Ramon Lull, who had lauded the chemistry of vegetable matter. "But," continues Libavius, "neither the transformation or adulteration of metals, nor the extraction of quintessences by means of the subtle spirit of wine, nor distillation, comprise the whole of the chemical art, although they are related."[22] Libavius detected a more comprehensive approach in the *Archidoxis* of Paracelsus, which he regarded as the latter's unsuccessful attempt to reduce chemistry to the form of an art. While Paracelsus had included elements of both the alchemical and distilling traditions, he had not set out the operations which underlay his preparations and had singularly failed to give a proper disposition of the art. As Libavius remarks contemptuously, how could such an utterly ignorant man be expected to accomplish something which depended upon a knowledge of logical method, which is the gateway to the constitution of the arts?[23] From this discussion we can discern Libavius's intention: it is to succeed where Paracelsus had failed, namely, to seek out the related operations and preparations separately described in the diverse chemical literature and to organize these logically into the form of a comprehensive art.

Two problems remain: What was this chemistry which Libavius

[20]Ibid.
[21]Ibid., 1: 112–13.
[22]Ibid., 1: 113.
[23]Ibid., 1: 114.

proposed? And how did he intend to organize it? The first of these is the subject of two letters in his chemical epistles (letters 7 and 8, "Quid Chymia" and "De Eadem Definitione").[24] He begins with some critical reflections on Paracelsian notions of chemistry. Everyone knows, he states, what *they* mean by the spagyric art: it is the separation of the pure from the impure. He recalls an uncomfortable journey through the Thuringian Forest in the company of one of them during which the sole topic of conversation was the separation of the pure from the impure. The Paracelsians seemed to imagine that whatever involved a cleansing of things was chemistry. Thus they believed that brickmakers (?) (*Molitores*), cooks, unguent makers, and millers were unwittingly practising the art. But if all these were admitted to the company of spagyrists, then, Libavius asks, why exclude the potters, the perfumers, the paintmakers, and the cosmetic manufacturers (*cerussandarum virginū adornatrices artes*)?[25] But whatever the status of these artisans, Libavius believed that a study of the writings and operations of the chemical arts would reveal the true object of chemistry. He enumerates some of these. It is concerned with extracting the most pure juices from things so as to render them more powerful in their virtue and effects. Sometimes it mixes substances together; other times it separates them; and it can also change and transmute things and alter their color. But all of this serves the one end: namely, that the substances so prepared according to the precepts of the art attain their most powerful virtue and emerge from their processing in a more noble form, having cast aside their useless impurities.[26] Such genuine chemical purifications do not, however, encompass the activities of the washerwoman (*lotrix amicula*) or the furrier cleaning his skins or the shoemaker tanning his hides. Nor does he believe that the smelting of metals is a proper component of the art, although he admits that the good chemist should be well informed on the metallurgical techniques and instruments. The same considerations apply to pharmacy.[27] What Libavius is seeking to do is to abstract chemistry from its applications in these various trades and skills and to set it out as an independent art. He finds it hard to formulate a suitable definition of this independent art. He suggests that it has to do with the dissolution of solid substances and mixtures, the cleansing of their dissolved parts, and their recomposition to produce the truest, most powerful, and most appropriate essences. But this is not wholly satisfactory, since chemistry does not

[24]Ibid., 1: 84–88 and 88–91, respectively.
[25]Ibid., 1:84.
[26]Ibid., 1: 85.
[27]Ibid., 1: 85–86.

always seek to combine substances; sometimes it simply transmutes them.[28] He has similar difficulties finding an appropriate name for the subject. Should it be called the art of fusing, the art of dissolving, or the art of distilling?[29] No one name or definition seems adequately to condense the various operations and functions of the art. One thing of which Libavius is certain, however: such an art exists, and it can be suitably organized. "If you wish to define chemistry," he states, "you will easily recognize the art: for it consists of homogeneous precepts set out in methodical form and order."[30] In the end Libavius seems to say that one cannot know what chemistry really is until its various operations and prescriptions have been properly set out and organized. Thus the definition must await methodization of the art. Chemistry, for Libavius, was the sum of its parts methodically arranged.

While these reflections might appear perplexing and inconclusive at first sight, throughout his discussion Libavius employs certain key terms which are the clue to his inspiration and intention. The significant words are *precept, invention, judgment, disposition, method,* and *art.* All of these were critical technical terms in sixteenth-century pedagogical theory, which in turn was based on theories of dialectic. As he himself states at the end of the epistle on the definition of chemistry, "Dialectic is the art of discovering simple notions and of unveiling them in order for judgment."[31] To understand Libavius's endeavor it is necessary to consider the central position of dialectic in the cultural tradition of the sixteenth century.

The Art of Arts

The most influential book in the history of Western thought from the thirteenth through the sixteenth centuries opened with the oft repeated sentences:

> Dialectic is the art of arts and the science of sciences, possessing the way to the principles of all curriculum subjects. For dialectic alone disputes with probability concerning the principles of all other arts, and thus dialectic must be the first science to be acquired.[32]

[28]Ibid., 1:86–87.

[29]Ibid., 1: 89–91.

[30]"Tu si definire chymiam velis; artem esse facilè agnoveris. Constat enim praeceptis homogeneis, methodica forma ordinéque dispositis" (ibid., 1: 84).

[31]Ibid., 1: 89.

[32]This translation is cited from Walter J. Ong, *Ramus: Method and the Decay of Dialogue* (Cambridge, Mass: Harvard University Press, 1958), p. 60. The

The book was the *Summulae Logicales* of Peter of Spain (Pope John XXI, d. 1277), a work which in its original and many derivative forms served as the primer for legions of students who passed through the universities of Europe during two and a half centuries. It was the rudimentary text of the arts curriculum and was intended to provide the neophyte with the basic intellectual tools for all his subsequent study. The work was nominally a compendium of logic, and yet it began with a paean to dialectic. The second sentence is even more curious. It states that dialectic disputes with *probability* but also that dialectic is the first *science* to be mastered. The apparent confusion here was a very real one, arising from the early scholastics' failure to delineate precisely the distinct modes of reasoning discussed in the six books of the logical corpus of Aristotle, collectively known as the Organon,[33] which passed down to the Middle Ages. The two modes were logic proper and dialectic. In a sense the confusion was understandable and well-intentioned. It arose in part from an effort to impose an intellectual unity on a corpus of texts, some of which were imperfect in themselves and which collectively were never intended to form a unified treatise on the reasoning process. Aristotle's distinction between logic and dialectic was rooted in the different styles of argumentation to which he was exposed in Plato's Academy. Dialectic had to do with the form of philosophical discourse manifest in the Socratic dialogue. It involved the searching out of suitable propositions and arguments and the ordering of these in such a way as to convince an adversary or neutral correspondent of a certain philosophical position. The discourse began with probable arguments and sought by dialogue to refine these to true conclusions. Logic, on the other hand, was concerned with scientific demonstration and had its roots in the mathematical tradition of the Academy. In his logic proper, the most important treatise of which is the *Posterior Analytics,* Aristotle sought to extend the rigorous resolutive and compositive methods of geometry into verbal modes of reasoning by means of the strictly controlled syllogism so as to bring the non-mathematical elements of physical reality within the domain of axiomatic science. The difficulty in distinguishing neatly between logic and dialectic in the context of the Organon is that the syllogism is common to both modes of reasoning.

The medieval scholastics of course came to recognize the subtle distinctions in Aristotle's logical corpus and exploited them to the full.

---

discussion of dialectic and method which follows is based largely on this work by Ong and on Neal W. Gilbert, *Renaissance Concepts of Method* (New York: Columbia University Press, 1960).

[33]Six books constituted Aristotle's Organon: *The Categories, On Interpretation, The Prior Analytics, The Posterior Analytics, The Topics,* and *The Sophistic Refutations.*

But these subtleties of reasoning were well beyond the requirements and capabilities of entering university students. The confusion between dialectic and logic continued to haunt the recesses of the European intellect through generations for centuries, with the most surprising consequences. It was this blurred edge which led Andreas Libavius to believe that he could invent the art of chemistry and pass it off as an axiomatic demonstrative science which was a branch of natural philosophy. Indeed it was this very aberration which made it possible for seventeenth-century Europeans to entertain the idea that systematically ordered knowledge was actually science.

Already in the thirteenth century, the balance between logic and dialectic was titled on the side of the latter. This is clearly visible in the opening of Peter of Spain's logical manual. This disequilibrium is explainable in part by the nature of medieval university pedagogy. While formal teaching was comprised of reading and commentary on the standard text, examination was conducted orally by means of scholarly dialectic. The defense of the proposition and the finding of arguments to persuade or defeat one's opponents were the mental skills which ensured a successful university career. These were in the main skills of dialectic and not of logic. The humanist movement of the fifteenth and sixteenth centuries, particularly as it was manifest in the centers of learning north of the Alps, did nothing to correct this imbalance. Indeed the scales were tilted more heavily on the side of dialectic by the renewed emphasis on eloquence, which threw in rhetoric to that scale of the balance. So heavily weighted were matters on the side of refined discourse that the distinction between logic and dialectic was virtually obliterated, and a new counterpoise of verbal skills was established between, on the one hand, dialectic, which was thought implicitly to include logic, and rhetoric on the other. Other external but related factors were at work in this reordering of the taxonomy of verbal skills. One was the humanist concern for pedagogical reform, which emphasized the speedy acquirement of the fundamental skills of correct and refined discourse, which then could be exploited for useful social ends, and not simply in intellectual dueling matches. The other was the advent and diffusion of the printed book. Out of this cauldron of pedagogical reform, social concern, and technological innovation emerged a new mode of discourse which was particularly tailored to the physical nature of the book but whose technical vocabulary remained that of an earlier oral culture. But most astonishing of all are the echoes that that terminology evokes in the twentieth century—art, science, method, invention, and even technology itself.[34]

[34]The displacement of a dialectic conceived in oral terms by one which

The humanist reforms of dialectic essentially exaggerated and codified themes and concepts which were both explicit and latent in medieval scholarly discourse. The humanists also found new meaning for concepts and terminology in the classical oratorical and rhetorical literature which they so favored and cultivated. Two concepts and terms came to have particular import in their pedagogical program, namely, art and method.[35] *Methodus* was a word of uncertain etymology and vague definition in medieval philosophy. Although the Greek term (μέθοδος) found a prominent place in both Plato (especially the *Phaedrus*) and Aristotle (the *Topics* and the *Physics*), the early medieval translators generally did not transliterate the Greek into Latin; instead a number of synonyms came into use, such as *via, ratio, processus, modus,* and *ars.*[36] The Latin equivalent *methodus* did have some currency from its use in Boethius's translation of the *Topics* of Aristotle, but its precise definition was obscured by its association with this host of other synonyms.[37] Not that there was a lack of interest in method in the Middle Ages. Aristotle spoke prominently about "method" in both the *Topics* and the *Physics;* but nowhere did he give a precise definition of what he meant by the term. For the most part the medievals took their cue from the *Topics,* where dialectic was described as "the path to the principles of all inquiries."[38] In its "topical" context Aristotle's method had to do with the arrangement of questions and arguments in proper series in dialectic. This dialectical setting of method also contained a pregnant idea which was to become an important part of the connotation of the term, namely, that method was a way or a road through knowledge.[39] This linked the process of "invention," the finding of arguments, with the disposition of arguments in dialogue. By some subtle transformation, whose stages are by no means clear, the concept of method became increasingly associated with the idea of a short-cut to knowledge. This is already manifest in the thirteenth century, when method is confounded to some degree with the presentation of an art, taken in its most general meaning as a curriculum subject, especially the concise and systematic presentation

---

was accommodated to the visual imagery of print is one of the major themes of Ong's *Ramus.*

[35]These themes are developed in Gilbert, *Renaissance Concepts of Method;* see especially pp. 67–115.

[36]Ibid., pp. 48–50.

[37]The passage is quoted in ibid., p. 50. Some doubt, however, has been raised as to whether the translation was in fact by Boethius; see ibid., p. 50, n. 17.

[38]Aristotle *Topica* 1. 2. 101^b1–4, as translated by W. A. Pickard-Cambridge in *The Works of Aristotle,* ed. W. D. Ross (London: Oxford University Press, 1928), vol. I.

[39]Gilbert, *Renaissance Concepts of Method,* pp. 58–59.

of an art in an introductory compendium such as Peter of Spain's *Summulae Logicales.* Already in the Middle Ages the terms "art," "method," and "brief way" have merged in a somewhat ill defined manner, most probably, as Ong suggests, out of the exigencies of the pedagogical pressures of the university arts curriculum.[40] In the introductory curriculum text we have come a long way from the anything but brief mode of philosophizing outlined in Aristotle's *Topics* and exemplified in Plato's Socratic dialogues. The concise textual presentation of a curriculum subject, however, never overwhelmed medieval pedagogy. The student had always to meet the final test of his dialectical skills in the oral *quaestiones* which constitued the examinations. It was only under the influence of fifteenth-century humanist pedagogical reform that method itself would be "methodized," given new prominence and more precise definition, and hustled in the back door to participate in genuine philosophical discourse.

Initially the early humanists avoided the Latin word *methodus* for reasons similar to, if more consciously fastidious than, those of early medieval translators—Cicero had not Latinized the Greek term but instead had used the expressions *via* and *ratio,* sometimes even conjoined.[41] This was an inhibition they soon overcame, however, and method became a central plank in their educational program. One of the important nuclei around which humanist notions of art, method, and pedagogy crystallized was the Stoic definition of an art, or *techne,* as a "system of percepts exercised together toward some useful end in life."[42] As the term "percepts" indicates, this definition, whose author was Zeno, was an epistemological statement. But by one of those curious stylographic inversions *perceptio* was transformed into *praeceptio,* thus completely obscuring the epistemological intent of the original.[43] Indeed the Stoic patrimony of this definition was not generally recognized. It was derived secondhand from a number of favorite humanist authors, including Cicero, Quintilian, Lucian, and the orator Hermogenes.[44] But in its epistemologically neutered form, this Stoic definition of a *techne* was to have powerful impact on humanist educational theory. It catered to all of the humanists' pedagogical proclivities by defining an art in terms of a set of precepts (teachings) exercised together (methodized) for some social purpose. In this form it

[40]Ong, *Ramus,* pp. 152–58.

[41]Gilbert, *Renaissance Concepts of Method,* p. 49.

[42]This important definition is discussed in ibid., pp. 11–12. Cf. Ong, *Ramus,* pp. 227–28.

[43]Gilbert, *Renaissance Concepts of Method,* p. 12.

[44]Ibid., pp. 69–70.

was irresistible, sanctioned as it was by such a galaxy of the masters of eloquence. The whole of the humanist educational program came together here. Method became the means by which curriculum subjects were systematically and concisely presented so that students could speedily master the principles and go on to apply the knowledge so acquired to useful social ends. All subjects were viewed as ripe for methodization, and methodized textbooks on a welter of arts appeared with increasing frequency from the printing presses of Europe throughout the sixteenth and seventeeth centuries.

The first important step towards providing humanism with a coherent form of pedagogical discourse, which was to become peculiarly suited to the printed book, was taken by Rudolph Agricola (1444–85), whose principal and most influential work was his *De inventione dialectica libri tres*, completed in 1479.[45] The focus of Agricola's dialectic was the topics, which under his influence became known as the *loci,* or places, a special category of which were the commonplaces. Good humanist as he was, Agricola turned for his sources on the topics to Cicero and Quintilian, as well as to Aristotle and Boethius. In so doing he confounded, not unconsciously, the two *loci* of topical invention, dialectic proper and rhetoric. Cicero and Quintilian had dealt with topics as they pertained to rhetoric, that is, in the context of finding arguments for discourse meant to persuade or move the auditor to action; Aristotle's *Topics,* on the other hand, was concerned with finding arguments for dialectic as philosophical discourse intended to convince intellectually. (Aristotle dealt with rhetorical topics separately in his *Art of Rhetoric.*) Agricola's tactic in merging dialectical and rhetorical invention was to produce a unified mode of eloquent discourse which would have a multipurpose function and become rhetoric simply by the addition of suitable ornamentation. The topics, or *loci,* were thus excised completely from rhetoric by Agricola and placed wholly within dialectic. Rhetoric was made to concern itself exclusively with appropriate flourishes and decorations. In fact Agricola's all-purpose dialectic had one overriding purpose, namely, teaching:

> The aim of speech can never be simply pleasure as distinct from moving and teaching another. For all speech teaches. . . . As far as invention goes, there is no difference between moving and teaching, for to move someone is to teach him as far as this is possible.[46]

[45] The following discussion of Agricola is based on Ong, *Ramus,* chap. 5, pp. 92–130.

[46] Translated and quoted in Ong, *Ramus,* p. 103.

Agricola's dominant mode of discourse subsumed logic entirely. The subtle distinctions between probable reasoning and scientific reasoning were largely irrelevant to him; it was sufficient that the subject matter in hand could be taught. Not that he was overtly anti-Aristotelian or adamantly opposed to the syllogism. Syllogistic argument simply did not figure prominently in his dialectical logic, and inasmuch as it was indulged as an end in itself, it was pedagogically worthless and hence to be shunned as a teaching or learning process.[47] But as Ong has pointed out, there was a much more significant development buried deep in Agricola's formulation of dialectic. By his overwhelming emphasis on the topics, or places, conceived of as a general and generic classification of topical headings which could provide something to say on almost any subject matter, Agricola was subconsciously eliminating much of the truly verbal and oral content of classical and medieval dialectic. The topics, especially in their new formulation as *loci,* readily formed the basis of a visual classification. A dialectic which did not pass much beyond the *loci* to the predicables of the *Categories* or to the syllogism was one that, wittingly or not, was most amenable to codification on the printed page.[48]

Agricola's topical logic was incomplete. Only the first three books on invention were ever published; the remainder presumably would have dealt with the second part of Ciceronian dialectic, namely, judgment, the arrangement or disposition of the arguments dislodged from the topical places. The subsequent history of dialectic up to the middle of the sixteenth century can be seen as an effort to supply this second part of Agricola's dialectic. Two figures in this development are important for our story—Melanchthon and Peter Ramus. Both in their separate ways represent the epitome of the humanist pedagogical tradition: the former is well known as the *praeceptor* of Lutheranism; the latter may justly be considered the *praeceptor* of Calvinism. Each of them responded to the topical logic of Agricola, but in quite different ways temperamentally. As a loyal Aristotelian, Melanchthon attempted to incorporate the new topical emphasis into a conventional Aristotelian presentation of logic which respected both the content and sequence of the Organon. Ramus, on the other hand, exaggerated the topical emphasis and made it the basis of a wholly new and distinctive dialectic, with its own rules of judgment and method.

In Melanchthon's principal dialectical work, the *Erotemata Dialectices*[49] of 1547, the oral residues of dialectic are still plainly visible.

47 Gilbert, *Renaissance Concepts of Method,* pp. 76–77.

48 Ong, *Ramus,* pp. 104–12.

49 For an analysis of the *Erotemata Dialectices,* see Gilbert, *Renaissance Concepts of Method,* pp. 125–28; and Ong, *Ramus,* pp. 236–39.

The first three books deal with the subject matter of Aristotle's *Categories* and part of the *Analytics,* namely, the predicaments and predicables, propositions, argumentation, and induction, which collectively represent the judicative part of dialectic in the Ciceronian division. Book 4 deals with invention and the *loci.* A section on method is squeezed in between Books 1 and 2 of the work. The whole organization of the work is patently an effort to accommodate the new humanist rhetoric-dialectic, with a fashionable concern for method, within the traditional structure of the Organon. But the new themes eat into the substance of the contents as well. Thus the predicables are strangely condensed into single words, or simple notions and themes housed in their topical abodes. They are to be dislodged, however, by means of appropriate questions. Melanchthon's method turns out to be nothing more than the positing of a standard ten questions of a single word or term under discussion. Four of these are taken from the *Posterior Analytics* of Aristotle, and the other six from the *Topics.*[50] Melanchthon's dialectic has a topical emphasis, but he labors to preserve remnants of all parts of Aristotle's logic, including the *Analytics,* thus attempting to preserve a "scientific" aura around the whole. Melanchthon's reputation and influence stemmed more from the use and example of his successful textbooks on the curriculum subjects than from any originality of his statements on dialectic and method.[51]

In contrast with the "gentle reformer," Melanchthon, Peter Ramus (1515–72) comes on stage like a whirlwind.[52] His aggressive reforms in dialectic, rhetoric, and method, which reached out into all of the curriculum subjects, were in part the forced flowering of the humanist topical logic in the more intense philosophical and pedagogical climate of arts scholasticism at the University of Paris. Ramism, as his collective reforms came to be known, launched the sixteenth-century educational world into a highly emotional debate to which no educator in northern Europe at least could remain indifferent. So powerful was the impact of Ramism that the historian instinctively seeks out the intellectual core of its challenge and appeal. But its essence proves difficult to grasp, and the philosophical insight which

[50]The questions are listed in Gilbert, *Renaissance Concepts of Method,* p. 126.

[51]Ibid., p. 127.

[52]The classic study of Ramus until recently was Charles Waddington, *Ramus: sa vie, ses écrits et ses opinions* (Paris: Charles Meyrueis et Cie, 1855). This has largely been superseded by Ong's study. Much can also be learned about Ramus and Ramism in Wilbur S. Howell, *Logic and Rhetoric in England, 1500–1700* (New York: Russell and Russell, 1961). An important study of Ramus' importance for science is R. Hooykaas, *Humanisme, Science et Réforme: Pierre de la Ramée (1515–72)* (Leyden: E. J. Brill, 1958). The discussion which follows relies heavily on Ong.

one feels must lie at the heart of such a powerful and influential movement seems strangely elusive. The truth is that there was no such insight. The essence of Ramism was its very triteness, from which stemmed its effectiveness. In the Ramist reforms all subtlety of thought, all philosophical niceties, are swept aside in the wake of pedagogical expediency. Its one great virtue—the original stimulus for its author—was that it made life a good deal simpler for the teachers of the world. Once diffused in the classroom, Ramism seeped into the consciousness of Western Europe before it was old enough to protest.

Method was the thing. Even Ramus's fabled anti-Aristotelianism turns out to have been, not a violent assault on the intellectual fortress of scholastic Aristotelianism, but the considerably less daring sally that Aristotle's Organon was confused and confusing and hence something less than an ideal teaching manual.[53] Ramus claimed to have penetrated the obfuscations of the Aristotelian corpus and to have encountered the true Aristotle. But he made the same claim concerning practically every philosopher of stature of the ancient world, including Socrates, Plato, Hippocrates, and Galen.[54] His own version of his insight was that he had recognized that all of these great thinkers and pedagogues (the two were synonymous for Ramus) had advocated one sole method of investigation and instruction. This turns out to have been, not surprisingly, the Ramist method itself.

It all began innocently enough in 1543 with Ramus's *Dialecticae institutiones,* which sought to supply the missing part of Agricola's text on the same subject. Indeed if it had not been for the author's *Aristotelicae animadversiones* published in the same year, Ramism might have slipped quite quietly into the world. Ramus's dialectic was wholly topical in its orientation and ignored completely the problems of predication. In the discussion of invention, an emaciated form of the syllogism survived. Arguments were to be generated out of the topics by the positing of explicitly formulated questions, which generally sought to link two terms. The question was mentally run through the topics—there were fourteen in the 1543 formulation—in search of the appropriate middle term, which then could be exploited in a simple syllogism. By this wholly automatic procedure subject matter for any type of discourse was invented.[55]

The truly innovative part of Ramus's dialectic, however, was the section on judgment, that part of dialectic which dealt with the arrangement and organization of what had been invented into con-

[53]Ong, *Ramus,* pp. 36–47.
[54]Ibid., p. 42.
[55]Ibid., pp. 172–95.

tinuous discourse. In this first formulation, Ramus spoke of three judgments. The first was the disposition of a single argument of invention in syllogistic form, as described above: it also included a discussion of enthymeme, induction, and example. The second judgment dealt with the concatenation of various arguments into a continuous discourse directed at a specific end. The third judgment was a strange theological entity which sought to refer everything to the light of God. This last was of no practical consequence and was most probably invoked, according to Ong's suggestion, to provide a Platonic gloss to Ramus's work, which allowed Ramus to distance himself even further from the Aristotelians.[56]

It was the second judgment which was of most significance; it was this which evolved into Ramus's method.[57] The process of linking arguments advanced by nothing more mysterious than the definition of terms and their subsequent division into further definables. Thus discourse appeared to evolve automatically out of itself by a self-replicating process of division and definition. Ramus vaunted the progressively linear character of his dialectic over what he regarded as the essentially circular mode of syllogistic reasoning. If this seems a somewhat banal and depressing method of generating a continuous statement about anything, it should be borne in mind that Ramus was here developing the protean subject of all curricular instruction, and the dominant objective was clear and tidy presentation.

In its subsequent formulations the essential character of the second judgment did not change; only the claims made for it became more expansive: it becomes *the* method. In Ramus's printed works, the definitions and divisions acquire the attributes of a more tidy classification system, and the characteristic bracketed dichotomies become more prominent. The method in its various revisions increasingly abrogated to itself the terminology and claims of an axiomatic system claiming to link cause to effect. Ramus claimed support for this in the methods of Plato, Galen, and even of Aristotle himself. In this last context, Ramus cunningly linked his divisions and subdivisions to the categories of genus and species without ever being quite clear on the distinction and, by exploiting Aristotle's comparison of the relationship between an axiom at the head of a series of deductive reasonings to a genus at the head of its species, passed off his bracketed flow charts as axiomatic deductive schemes of reasoning. Ramus's purpose, however, was to be pedagogically convenient, not philosophically illuminating or even con-

[56]On Ramus's third judgment, see ibid., pp. 189–90.

[57]On Ramus's method, see ibid., pp. 225–69; and Gilbert, *Renaissance Concepts of Method*, pp. 129–44.

sistent. As he puts it in an early statement (1546) of his method: "The method of teaching, therefore, is the arrangement of various things brought down from universal and general principles to the underlying singular parts. by which arrangement the whole matter can be more easily taught and apprehended."[58] Such pedagogical method, for all its philosophical naiveté, had several distinct advantages: it was universally applicable; it was consistent and straightforward in its procedures; it could be conveniently summarized in diagrammatic form on the printed page; and it was an effective memory device.[59]

Libavius and Ramism

As a schoolmaster, Andreas Libavius was inevitably drawn into the great debate about method and pedagogy which Ramism precipitated. Indeed his first polemical work (published in 1591) engaged two Ramist natural philosophers, William Temple, of Cambridge, and James Martin, a native of Dunkeld in Scotland, who was a professor of philosophy at the University of Turin.[60] At first sight this work might appear to place Libavius firmly in the anti-Ramist camp, but matters are not so simple. The principal points at issue in this polemic were not Ramist reforms in dialectic or method; rather they were concerned with natural philosophy. Martin had written a treatise which attacked the Peripatetic doctrine of the generation of material species from the four elements and also the related doctrine of the four humors of Galen, to which Temple had added an introductory preface.[61] In response Libavius had risen to the defense of orthodox Peripatetic element theory, entitling his reply "A Treatise on Physical Questions of Controversy between the Peripatetics and the Ramists." While the title serves to highlight Libavius's Aristotelian orthodoxy and correctly identifies

[58]Quoted in Ong, *Ramus,* p. 245.

[59]Ramist method considered as a memory device is treated in Frances A. Yates, *The Art of Memory* (Harmondsworth: Penguin Books, 1969), pp. 228–38.

[60]*Quaestionum Physicarum, Controversarum inter Peripateticos et Rameos, Tractatus; in quo disceptantur octo quaestiones, ex illis quae de Elementis nuper ut inaudita protulet contra Aristotelem Iacobus Martinus Scotus Dunkeldensis, Philosophiae apud Taurinenses in Anglia* (sic) *professor publicus. Cum Praefatione Guilielmi Tempelli, Cantabrigiae philosophi; quae itidem ad veritatis normam exigitur . . . ab Andrea Libavio* (Frankfort, 1591).

[61]*Iacobi Martini Scoti Dunkeldensis, Philosophiae Professoris Publici in Academia Taurinensi, de prima simplicium, & concretorum corporum generatione, Disputatio, in qua Aristotelis, Galeni & aliorum sententia de simplici & absoluta generationem desiderentur, proponitur . . . Cum Praefatione Guilielmi Tempelli Cantabrigiensis . . .* (Frankfort, 1589).

his opponents as two followers of Ramus, his response to Temple's preface in particular reveals a somewhat more ambiguous attitude to the phenomenon of Ramism itself.[62]

In this reply we again find Libavius's hostility to enthusiasms and his pedagogical concerns to the fore. He notes the raging controversies concerning method which have convulsed the age.[63] These he deprecates for two reasons: firstly, because they have consumed the energies and time of scholars on this one topic, so that progress in philosophy has been frustrated; and secondly, because they have had a disastrous effect on the education of youth, who are seen to defend absurd theses drawn from authors whom they do not understand, and with nothing but the worst sort of sophistry.[64] Libavius feigns sympathy here with the students who have to decide between the two principal factions in logic, the Aristotelians and the Ramists.[65] The pity of it is that the whole controversy is unnecessary, since both sides are striving towards the same goal: he compares it to the day-long debate the traveler has with himself about which of two ways to take to the same place—both may be equally safe, convenient, and short.[66] Libavius concedes that many very able men have found the Ramist method acceptable and that there are some who think that his (Ramus's) terms and method of setting things out are extremely learned and clever.[67] The chief problem of Ramism, as Libavius sees it, is the deadly strife and fury which it has provoked. He has no doubt where the responsibility for this sorry state of affairs lies: it is with the Ramists, beginning with Ramus himself. While the latter was able to demonstrate something of value in his exposition, he negated it by his refusal to entertain improvements in his program and by his biting criticism of his adversaries. The result has been that the tumult which he engendered has disturbed everyone, and thus his valuable book on

62*Ad Lectorem de Praefatione Tempelli contra Aristotelem,* which follows the *Dedicatio* of the *Quaestionum Physicarum Tractatus,* unpaginated (hereinafter to be cited as *De Praefatione Tempelli* with page numbers supplied).

63"De artium constituendarum forma, quantis viribus decertatum sit, & adhuc inter sese digladientur ij qui eruditorum nomen sibi arrogant, non est obscurum" (*De Praefatione Tempelli,* p. [1]).

64Ibid., pp. [1–2].

65"Verum cum duae nostris temporibus sint potissimum Logicorum factiones, Aristotelea & Ramea; inter eos quorum fidei commissa sunt tenera ingenia, cōstare debebat, utri potissimum sit adhibendus adiugendusque discipulus" (ibid., p. [2]).

66Ibid., pp. [2–3].

67"Sed & Ramea Methodus ingeniosissimis non displicet; inveniunturque qui terminos & modum descriptionis eius pro valdè eruditis habent & felicibus" (ibid., p. [3]).

logic has been banned from many schools. Our schoolmaster, ever ready to thrust home the moral, speculates that things might have gone differently for Ramus if he had set aside his scornful fury and evoked the sympathy rather than the hatred of knowledgeable men.[68] (It is as if Libavius were dressing down a bright but incorrigibly fractious pupil.) Alas, the followers of Ramus have proved no better than their mentor: instead of restricting themselves to the task of providing appropriate arrangements for those disciplines which hitherto have been badly set out, they too have preferred to wound the reputation of others with outrageous and abusive language.[69]

Both the substance and the tone of this Libavian critique of Ramus and Ramism are now familiar to us. It depicts Ramism as an affliction visited upon the conscientious schoolmasters of the world. Its associated stridency has disturbed the peaceable and progressive conduct of the academic life: it has sown doubts in the minds of schoolboys, so that they do not know whether to trust their teachers or not; and it has completely ruined their exercises! The source of all this trouble was quite simply explained by Libavius. Ramus and his followers were unreasonable men. They showed no respect for the differing views of their predecessors or contemporaries. They were know-alls. They preferred invective to constructive dialogue. In short, like the Paracelsians, they were enthusiasts.

The Ramists, however, were enthusiasts of a different stamp from the Paracelsians. They were unreasonable men who sought reasonable goals; not irrational men who strove to grasp the supra-rational. Unlike the Paracelsians, they were not beyond redemption, for dialogue could be established with them if only they would cooperate. In spite of his instinctive revulsion against the turmoil which Ramism had produced, Libavius clearly was willing to entertain the proposition that its reforms had certain merits. If only the Ramists could curb their strident tone, perhaps they could make a positive contribution to scholarly discourse. The contribution of Ramism which appealed to Libavius was obvious even in this early polemic: it was the method which it provided for the disposing of the arts, particularly those arts which lacked methodical presentation. Libavius's position vis-à-vis Ramist dialectic and method can be more directly estimated from his own texts on dialectic. The first of these was published in 1593 for his

68Ibid., pp. [3–4].

69"Facerēt ita hodie illi, qui Ramum non tam in arte enucleanda, quam latrando & aliorū scripta evertēdo sequuntur; & potius excolere ipsi scientias maleque formatas forma propria exornare, quam improbis maledictis famam aliorum lacerare mallent" (ibid., p. [4]).

students at Rothenburg; two supplementary texts appeared in 1595, the same year as the first two volumes of chemical epistles.[70] His definitive textbook on dialectic with a companion section on rhetoric appeared in 1608, when he was director of the Gymnasium Casimirianium in Coburg.[71] All of these texts, even in their titles, show Libavius to have been one of the self-styled Philippo-Ramists, who sought to combine Melanchthon's dialectic with that of Ramus.[72]

Philippo-Ramism was a posture quite widely adopted in Lutheran Germany. It was in the main a defensive position taken in the face of the sweeping tide of Ramism, which had come to be intimately tied to Calvinist education and culture. Ramus after all had been a genuine Calvinist martyr in the Massacre of St. Bartholomew's Day (1572). The Lutheran rear-guard action yielded on all essential points but sought to save appearances by preserving the memory of Melanchthon and his loyalty to Aristotle. The Philippo-Ramists tried to preserve something of the structure of the Organon in their dialectical manuals, and they tried even harder than Ramus had done to integrate Aristotle's analytic logic into the new dialectic. The religious overtones of the Ramist debate in Germany are quite visible in Libavius's *Dialectica Philippo-Ramaea* of 1608. In the dedication to the consuls of Gotha he makes much of the assertion that perverted dialectics keep cropping up, leading in turn to heresies. He implies that Ramus only completed what Melanchthon had begun but could not finish due to the pressure of affairs, namely, to supply an effective method of disposition for instruction and learning. Significantly, Libavius advocates his text for use in schools adhering to the "pure confession of Augsburg," the doctrinal *ne plus ultra* of orthodox Lutheranism in the years of confessional strife at the turn of the seventeenth century.[73] Libavius's Philippo-Ramism was of a part with his opposition to Croll's Calvinist Paracel-

[70]The titles of these works, none of which I have seen, are listed in *Das Corpus Libavianum* (items 2, 6, and 7) in the modern German translation of Libavius's *Alchemia, Die Alchemie des Andreas Libavius: ein Lehrbuch der Chemie aus dem Jahre 1597*, trans. with illustrations and commentary under the supervision of Friedemann Rex, for the Gmelin Institute for Inorganic Chemistry, in association with the Society of German Chemists (Weinheim: Verlag Chemie, 1964).

[71]*Dialectica Philippo-Ramaea, ex descriptionibus et Commentariis Philippi Melanchthonis & Petri Rami . . . Addita est Rhetorica, Descriptionis Audomari Thalaei . . . itidemque Melanchthonianis Oratoriae praeceptis aucta* (Frankfort, 1608).

[72]For a discussion of the Philippo-Ramists see Wilhelm Risse, *Die Logik in der Neuzeit*, 2 vols. (Stuttgart-Bad Cannstatt: Friedrich Frommann Verlag, 1964), 1: 122–200. See also Ong, *Ramus*, pp. 297–301; Gilbert, *Renaissance Concepts of Method*, pp. 213–20; and Howell, *Logic and Rhetoric in England*, pp. 282–317.

[73]See Libavius's *Dedicatio* to his *Dialectica Philippo-Ramaea*.

sianism: both were efforts to preserve Lutheran humanism in the face of strident intellectual and theological doctrines. That Libavius should employ the tools of enthusiastic Calvinist intellectualism (Ramism) to spike enthusiastic Calvinist Paracelsianism and so produce the discipline of chemistry is surely one of the more curious episodes in the history of science.

In the dedication of the chemical epistles Libavius wrote about chemistry:

> The more I turn it over the more it pleases me, so much so that I have become of a mind to arrange its precepts in the form of the perfect sciences and to make it clear by means of the Aristotelian disposition of the arts and by the Ramist method.[74]

What this implied was discussed in full in one of the letters, entitled "The Method of constituting arts," which was ostensibly an invitation to the respondent, Zacharias Brendel, a professor of medicine at Jena, to construct the art of chemistry.[75] Libavius makes immediate reference to the disputes about method involving the Ramists, and he warns Brendel that he may become embroiled in these. While he does not wish to anticipate Brendel's decision on these matters, he does stress the necessity of *a* method, if only to put an end to those ill digested and wholly unmethodical texts on the arts, chemistry included, which were flooding the market. Libavius claims that the only true method is that set out by Aristotle in the *Posterior Analytics;*[76] but as becomes patently clear in the course of the letter, he is only supplying an Aristotelian logical gloss for perfectly orthodox Ramist views. This he accomplishes by more outrageous misrepresentations and misapplications of Aristotle than even Ramus himself had used. One major aim of all this is to demonstrate that the invented art of chemistry methodically set out would be a true demonstrative Aristotelian science in conformity with the canons of the *Posterior Analytics.* The first such sleight of hand occurs when Libavius states that art (*techne*) and scientific knowledge (*episteme*) do not differ in substance or in essence.[77] This is an obvious misrepresentation of a passage at the end of

[74]"Sed quo amplius eam verso, eò arridet magis adeo, ut ad formam perfectarum scientiarum accomōdare praecepta eius, & Aristotelea artium constitutione methodoque Ramaea illustrare animum induxerim" (*Rerum Chymicarum Epistolica,* vol. 1, *Dedicatio,* sig. 5 recto).

[75]*Rerum Chymicarum Epistolica,* 1: 116–24.

[76]"Sed omissis talibus, ipse iudico non discendendum esse à vetusta artium et scientiarum constituendarum methodo quam accuratissimè prosecutus est in posterioribus analyticis philosophus" (ibid., 1: 118).

[77]Ibid.

the *Posterior Analytics,*[78] where Aristotle is trying to deal with the problem of how the mind arrives at the axioms which are the starting point for any demonstrative science. Aristotle's argument devolves on a description of the process of induction, by which the memory retains the universal derived from particular sense-perceptions to form experience. To accent the primacy of experience, as opposed to the innate ideas of Plato, Aristotle states that it is from experience that the skill (*techne*) of the craftsman and the knowledge (*episteme*) of the man of science originate. As he makes quite explicit, however, *techne* exists in the realm of coming to be and hence cannot be scientific, whereas *episteme* exists in the realm of being. Libavius completely glosses over this fundamental distinction, claiming that an art is just a concatenation of scientific axioms applied to the realm of becoming so that a particular task is accomplished.[79] Hence chemistry methodically organized is both art and science. One further convenient blurring of terms is manifest here, namely, the identification of an art conceived as a properly organized curriculum subject with a *techne* conceived as a craft skill—a transition mediated by the Stoic definition of an art discussed above. Ramism flourished and made much of its impact through just such aberrations of Aristotelian terminology. Libavius also exploited them to the full.

Libavius utters with heartfelt piety the Aristotelian strictures about experience being the basis of scientific knowledge and the conditions necessary for a true scientific axiom.[80] It is fascinating to witness his ingenuity in applying the rules for scientific propositions and syllogisms to his conception of the topically arranged and methodized art of chemistry which he envisages in a Ramist scheme. The condition of homogeneity or catholicity had to be met.[81] Ramus, too, had invoked this condition, calling it his second law of method.[82] In its original Aristotelian context it set a condition for the middle term of a syllogism which stipulated that the predicate apply to the subject essentially. In Libavius's hands this boils down to the assertion that the art of chemistry should contain nothing which is not chemistry:

Thus the philosopher is required to join together those things which are homoge-

[78] Aristotle *Posterior Analytics* 2. 19. 100^{a}6–9, in *The Works of Aristotle,* ed. W. D. Ross, vol. 1.

[79] *Rerum Chymicarum Epistolica,* 1: 118.

[80] Ibid., 1: 118–19.

[81] Ibid., 1: 119.

[82] Cf. Ong, *Ramus,* p. 259; and Howell, *Logic and Rhetoric in England,* pp. 149–52.

neous according to the law of homogeneity, having excluded all that is heterogeneous. According to this doctrine nothing is to be admitted in chemistry, which is not chemical: nothing unless it be true: for everything false is heterogeneous.[83]

But even more daringly Libavius hauls in the famous four questions, which Aristotle, at the beginning of book 2 of the *Posterior Analytics,* had posited as the source of all scientific knowledge.[84] Again in their original context, these questions related to predicates of subjects and were meant to generate valid middle terms for scientific syllogisms. The first question which Libavius introduces—it is Aristotle's third—is, Does the subject predicated exist? By reasoning which I am unable to fathom, Libavius takes this question to refer to the general definition of the subject of the art. Here he is in real trouble, for this is one thing which he cannot provide for chemistry. In conventional Ramist wisdom, the subject of an art could be reduced to one term or activity. Thus arithmetic referred to number, geometry to measurement, music to harmony, and dialectic to discoursing. The typical Ramist definition then took the form: arithmetic is the art of numbering well; geometry is the art of measuring well; dialectic is the art of discoursing well, etc. Libavius was unable to come up with a unique activity or subject which succinctly defines chemistry. Chemistry has to be defined complexly, and all its operations set out methodically by definition and division, before its essence can be perceived. Hence Aristotle's question can only be asked *after* chemistry has been topically methodized.[85]

The other three questions—Is the connection of an attribute with a subject a fact? What is the reason for the connection? What is the nature of the subject?—are likewise all drawn into the discussion. Again Libavius seeks to apply these questions to the definitions and disposition of the art. Exactly how they are meant to apply is not very clear, but the strategy of suggesting that they do is patent. These questions were intended originally to provoke respondent syllogisms which related terms causally and hence scientifically. By suggesting that the definitions and subdivisions of his methodically arranged art answer to these questions, Libavius is implicitly asserting that a properly methodized art is in fact a shorthand concatenation of scientific syllogisms causally connected. By this strategy, it is suggested that the definitions of the terms of the art are its material cause, while their disposition is its formal cause.[86] Here Libavius makes much use of the

83*Rerum Chymicarum Epistolica,* 1: 119.
84Ibid., 1: 120–21. Cf. Aristotle *Posterior Analytics* 2. 1. 89^{b}21–25.
85*Rerum Chymicarum Epistolica,* 1: 120.
86Ibid., 1: 121–22.

genus-species relationship which Ramus himself suggested underlay his bracketed divisions. "There is nothing easier," Libavius writes, "than to express well the essences of divisions by the good definition. For the genus always contains the species, and manifests its nature."[87] The letter to Brendel ends with rhetorical inducements to undertake the methodization of chemistry. "Beginners will owe you everything. No longer will they wander in the vastness of solitude, but in safety they will perceive and lay hold of every portion of the art as readily as in a synopsis. What could be more dextrous in learning?"[88] As always the priorities and concerns of pedagogy reign supreme.

The Rhetoric and Dialectic of Chemistry

For Libavius the art of chemistry systematically set out was in a curious way coincidental with the practical operations and products of chemistry. He conceived of discourse about chemistry as in some way identical with the doing of chemistry. Thus in the methodized art the definitions are homogeneous; but chemistry prepares homogeneous species. So too does the proper disposition of the art express its essence (*exprimere suam essentiam*), while chemistry expresses essences from material species. It was this particular coincidence of the vocabulary of dialectic with operations and objects in the real world that made Libavius's formulation of chemistry on a dialectical model possible. In this dialectical epistemology, the word defined and expressed reality so precisely that it could move freely between the mind, the page, and the object without distortion. The world, and all activity in it, could and must be reduced to words. This coincidence of language and reality was not fortuitous: it was rooted in Ramus's own peculiar form of linguistic empiricism.[89]

Ramus believed that all man's knowledge had its origins in practice. Thus men spoke before they knew the art of discoursing; counted before they had mastered arithmetic; and measured before they understood geometry. These ideas were most fully developed by

[87] Ibid., 1: 123.

[88] "Tyrones tibi omnia debebunt: Iam enim non amplius in vastissima errabunt solitudine, sed ob tutum facili tanquā in synopsi omnia artis membra perspicient, arripientque, quo quid in discendo est dexterius?" (ibid., 1: 124).

[89] This aspect of Ramism is best discussed in Hooykaas, *Humanisme, Science et Réforme*, esp. pp. 20–32, 51–59, and 91–96.

Ramus within the context of dialectic and rhetoric, but as Hooykaas in particular has shown, there is ample evidence that he wished to extend them into the area of the relationship between the crafts and the sciences.[90] After the teachers, Ramus celebrated the doers of the world. Thus he conceived of his art of dialectic simply as the efficient ordering and defining of the rules of natural speech. These rules were of course most clearly perceived in the best speech, namely, the rhetoric of the great orators of antiquity. But they were the same rules that governed all discourse, even the speech of the child. Art simply organized them so that they could be more readily taught, exercised well, and put into practice: this in part rationalized the merging of dialectic and rhetoric. There was basically only one kind of discourse, and that was natural discourse. Ramus's dialectic was simply natural discourse methodized. All human activity, both individual and collective, according to Ramus, had three stages of development: first, man's natural faculty of reasoning or doing, derived by learning from experience; second, learning and instruction through art, which taught the precepts of reasoning and doing well; and third, exercise, which put into practice the precepts of art. In other words, art was nature methodized.

These ideas were of fundamental importance to Libavius, who was to take the empiricism latent in them and carry it beyond the realm of purely literary application into the real world of experience and action. In the crucial sentence of his chemical epistles he writes: "I will demonstrate, if God will assist me, that the operations of the true chemistry are the statements (*documenta*) of its own nature."[91] The idea which informed his major contribution to chemistry was that out of the cumulated rhetoric and experience of chemical skills and trades a dialectic of chemical techniques and preparations could be invented and methodized.

The Alchemia

The fruit of Libavius's labor was the *Alchemia*, published at Frankfort in 1597.[92] This work is the first text which conceives of

90Ibid., pp. 92–96 and passim.

91"Demonstrabo . . . si me Dominus adiuverit, quam omnes operationes verae chymiae sint ipsius naturae documenta" (*Rerum Chymicarum Epistolica*, 1: 44).

92*D.O.M.A. Alchemia Andreae Libavii Med. D. Poet. Physici Rotemburg. operâ e dispersis passim optimorum autorum, veterum et recentium exemplis potissimum, tum etiam praeceptis quibusdam operosè collecta, adhibitisq; ratione*

chemistry as an independent and integral discipline divorced from its applications and which seeks to organize the techniques and prescriptions of the subject in such a way that they can be taught. As such, it is much more than a late sixteenth-century encyclopedia of the operations and recipes of the various chemical arts. It marks the appearance of a new scholarly discipline on the intellectual horizon—the discipline of chemistry.

The motives and objectives which prompted this singular achievement are reiterated by Libavius in his *Praefatio ad Lectorem.* [93] This is composed with a studied modesty, intended to contrast with the bombastic rhetoric of the Paracelsians. The pedagogical incentive and inspiration for the work take pride of place. In the first sentence of the preface, Libavius states that he has gathered together in this one work the precepts about Alchemy to be found in various authors, which, on the basis of teaching practice, he has organized with the method which ought to govern the disposition of the sciences. This he has done in order to promote this branch of knowledge in the education of youth.[94]

The purpose of this pedagogical exercise was to establish chemistry as a scholarly discipline which could be "methodically" taught in schools, thereby bringing it into the public domain by making it a part of scholarly discourse. With the operations and prescriptions of the art clearly set out, society would no longer be held hostage to the frauds, deceits, and secretiveness of the Paracelsians and their ilk. A properly methodized art of chemistry would provide the necessary "judgment" by which all chemical preparations could be assessed.[95]

It is this laying out of the art in methodical form which Libavius regards as his principal accomplishment. "Some will demand of me my own experiments," he writes, "not those of other practitioners, which fill numerous other books. However, I do not teach my own art, but rather I set out the chemical art as it has been confirmed by the work of its practitioners. It is enough if the exposition and method of teaching are my own."[96] Libavius, however, is reticent

---

et experientia, quanta potuit esse, methodo accuratâ explicata, et in integrum corpus redacta (Frankfort, 1597).

93 Eight pages unpaginated, in the *Alchemia.* In citations following, page numbers will be supplied.

94 "E variis autoribus, & usu artium magistro, ductu methodi scientiis informandis attributae, lector benevole, ALCHEMIAE praecepta in unum opus congessi, ad quod ipsum tum alia me impulerunt multa, tum potissimum ut studiis iuventutis hac quoque sciendi parte prodessem" (ibid., p. [1]).

95 See esp. ibid., pp. [1–2].

96 "Nonnulli requirent à me mea experimenta, non aliorum artificum, quorum pleni sunt libri plurimi. Ego verò non meam artem doceo, sed artem

about his method in the preface to the *Alchemia*. He only broaches the subject in response to anticipated criticisms of it, where he seems to be markedly on the defensive. One of the objections to his text which he foresees concerns its basic organization. Why has he not divided chemistry into sections on pharmacy and metallurgy, with perhaps additional parts on the analysis of mineral waters and metallurgical assay?[97] In response to this he replies that he had thought long and hard about the proper disposition of the art and had frequently changed his mind on the matter; but in the end the nature of the method unfolded itself, and he had only to follow where it led.[98] Thus when an operation was common to both pharmacy and metallurgy it should only be set down once. The same principle should apply to all the preparations and operations of chemistry, irrespective of their origins or applications in a particular chemical craft. A treatment of assay and mineral water analysis is deferred to the later commentaries on the *Alchemia,* as these techniques also employ precepts from other arts and are therefore synthetic and not homogeneous.[99] The method of the *Alchemia* is thus a "natural" method, in which the elements evolve out of themselves, and which is entirely homogeneous. These are, of course, characteristics of a Ramist art.

Libavius's *Alchemia* is in fact a thoroughgoing exemplar of a systematic textbook employing Ramist principles. Its contents evolve by means of definitions, divisions, and subsequent definitions of the subject matter proceeding from the general to the specific. These divisions are summarized at the beginning of the work in the form of bracketed tables (see the plates preceding chapter 4 of this volume), which set out visually the relationship of the various techniques and operations of the art.[100] The text itself is a statement of the definitions and an elaboration of the divisions illustrated by examples, so constructed that the discourse about chemistry seems to evolve naturally out of itself, giving the appearance of a logically coherent whole which proceeds from genus to species and, by implication, from cause to effect.

In spite of his text's obvious Ramist inspiration, Libavius remained in the end reluctant to acknowledge this. At the close of the

---

Chymicam usu artificum comprobatam, expono. Si mea est expositio & modus docendi, sat est" (ibid. p. [4]).

97Ibid., p. [6].

98"Cogitavi & ipse sedulò de constitutione apta: saepe mutavi cogitata probataq; prius. Natura tandem methodi seipsam explicavit: eius ductus fuit sequendus" (ibid., p. [6]).

99Ibid., p. [6].

100These tables follow immediately after the *Praefatio ad Lectorem*.

preface he somewhat defensively states that many of a "prickly temper" will think that the divisions and definitions of his method are comical. But, as he admonishes sternly, it is easy to criticize but difficult to supply something better. He concludes by looking forward to the day when the "silly chatter" of Ramism will become the exclusive concern of cynics.[101] Libavius's concern to divorce the application of essentially Ramist ideas from the overheated polemic about Ramism itself was not atypical. As Ong has commented, Ramism was a phenomenon which would swallow its own origins.[102] Libavius can be seen as an active participant in that process. But however much he wished to disown the fact, his major accomplishment of setting out the art of chemistry for the first time was dependent on the insight and stimulus which Ramism provided.

The various authors whom Libavius has consulted as sources for his dialectic of chemistry are meticulously listed in the pages following his *Praefatio ad Lectorem.* The list is highly eclectic and ranges in time from classical antiquity to Libavius's own contemporaries. From this list it is obvious that Libavius sought material from all of the major traditions which contributed to the elaboration of chemical technique and prescription: classical and Arabic writers on materia medica, medieval and Renaissance authors on distilling, alchemists, metallurgists, and Paracelsians.[103] The inclusion of the last group indicates that Libavius did not disdain the writings of the Paracelsians as providing examples of the chemical art;[104] he only challenged their self-asserted monopoly of

101"Methodi ratio divisiones, definitiones aliaq; procul dubio spinosis ingeniis oblectamenta parient mirifica. Facile est reprehendere, meliora dare difficile. Illud vel stultis licet, hoc sapientibus. Erit mihi sapiens, qui meliora exhibebit. Ramisticas argutias ineptasq; nugas ineptiunt hodie multi, ad Cynosarges abire iubeo" (*Praefatio ad Lectorem,* p. [8]).

102Ong, *Ramus,* p. 295.

103Classical and Arabic medicine are represented by Hippocrates, Galen, Dioscorides, Avicenna, Mesue, and Rhazes; the authors in the distilling tradition include Ramon Lull, Arnald of Villanova, John of Rupescissa, Hieronymus Brunschwig, Mathiolus, Philip Ulstadt, Conrad Gesner, Valerius Cordus, Adam Lonicer, and Walter Ryff; the Hermetic and natural magic tradition authors listed include Hermes, Roger Bacon, Giovanni Baptista Porta, and John Dee; the Paracelsians are represented by Paracelsus himself, Gerard Dorn, Leonard Thurneiser, Joseph Duchesne (Quercetanus), Thomas Muffet, and Thomas Jordan; Albertus Magnus and Geber represent the medieval alchemical tradition, while Georgius Agricola is the most prominent of the writers on metallurgy.

104"Non quidem repudiavi si quas formulas bonas apud Paracelsum inveni, quarum fors ipse autor non est" (*Praefatio ad Lectorem,* p. [6]). Later on Libavius vigorously repudiates the idea that chemistry was "invented" by Paracelsus: "Misera foret chymia, si ex Paracelso esset instituenda. Ego non tantum Paracelsica, & nostrorum cum veterum, tum recentium edita volumina & operas consului . . ." (ibid., p. [7]).

chemical knowledge and wisdom, their magical beliefs, their obscurity, and their anti-social behavior. In addition to these authors of printed works, Libavius appends the names of friends and correspondents with whom he has discussed chemical matters.[105] All in all, over one hundred individuals are listed. This fastidious acknowledgment of sources is intended to contrast with Paracelsian secretiveness and to show that progress in chemistry depends on the collective endeavor of candid practitioners making public their knowledge.

In terms of the dialectical analogy, these authors provided the rhetorical statements from which Libavius invented and methodized the basic dialectic of chemistry. To these should be added the *documenta* of Libavius's own personal experience. Despite his disclaimer that he did not teach his own art, Libavius does acknowledge including procedures and preparations of his own. These, however, are merged unobtrusively into the collective dialectic.[106] They only serve to underline the fact that in Libavius's world view it was only relevant that experience could be verbalized. His task, as he saw it, was to capture the verbal units of collective experience which flitted between nature and the text, to pin these down by definition, and to order them neatly on the page to produce a methodized art.

As every methodized textbook must, the *Alchemia* begins with the definition of the art. Libavius offers the following:

> Alchemy is the art of perfecting magisteries, and of extracting pure essences from compounds by separating them from their corporeal matrix.[107]

This is scarcely an unequivocal definition, and its meaning must await further elaboration of its terms. It is, however, Libavius's pragmatic solution to the problem which so vexed him in the *Rerum Chymicarum Epistolica*. Having got the preliminary requirement of definition out of the way, he quickly passes on to the primary division of the discipline into techniques and operations, or *encheria,* and the chemical species prepared by means of these operations, called simply *chymia.*[108] The dispositions of these two parts of *Alchemia,* which are

[105] One name which catches the eye in this list is that of the astronomer Tycho Brahe.

[106] "Multa addidi tamen etiam ex meo penu, quae alibi non invenies" (*Praefatio ad Lectorem,* p. [4]).

[107] "Alchemia est ars perficiendi magisteria, & essentias puras è mistis separato corpore, extrahendi" (*Alchemia,* p. 1).

[108] "Alchemiae partes sunt duae: Encheria & Chymia. Encheria est prima pars Alchemiae, de operationibus modis" (ibid., p. 2). Chymia is later defined as follows: "Chymia est pars secunda Alchemiae, de speciebus Chymicis conficiendis. Species Chymica est, quae per operationes Alchemiae in enchirisi expositas perficitur" (ibid., p. 85).

dealt with in separate books of the text, are summarized in the two bracketed flow charts at the beginning of the work. In effect these two tables most fully represent what Libavius believed chemistry to be.

The main part of the text of the first book of the *Alchemia* elaborates through definition and description the manual operations of chemistry depicted in the table of *encheria* (see the plate preceding chapter IV of this volume). Libavius, however, first devotes several chapters to a description of the instruments of chemistry which these operations employ. This propaedeutic of *encheria* includes a discussion of vessels, furnaces, and the control of fire for heating (*pyronomia*).[109] This also is defined and divided in a systematic manner. The principal category of instruments is vessels (*vasa*), which are defined as instruments suitable for holding things.[110] They are differentiated in terms of their material constitution, their capacity and shape, and their function. The various types of furnaces form a subcategory of vessels, since they are viewed as instruments for holding materials to be heated.[111] Their subdivision follows that developed for vessels. While Libavius attempts to bring order to containers and furnaces in this way, he incorporates all other utensils into one catchall category, called the *supellex tumultuaria* (literally "the disordered utensils"). These apparently defied his capacities for methodization, and he was content to list them in one single chapter. They include such items as filters, funnels, tongs, presses, bellows, condensers, and even spectacles for protecting the eyes from acrid vapors.[112]

The whole range of chemical operations are spiked on a complex series of dichotomies. The techniques of elaboration (*elaboratio*) operate on the material substance,[113] while those of exaltation (*exaltatio*) produce changes of form and quality.[114] The former in turn is dichotomized into those operations which break down the solid structure and continuity of bodies (*solutio*) and those which solidify bodies (*coadunatio*). *Solutio* in turn may resolve a body into a homogeneous fluid whole either by heat, as in fusion, or by the absorption of moisture, as in deliquescence. Alternatively the body may be resolved into heterogeneous parts which in turn can be segregated (*segregatio*)

109Ibid., pp. 2–28.

110"Vasa sunt instrumenta capiendis rebus idonea. Itaque & capacitatem habent in se, & sua orificia" (ibid., p. 3).

111"Fornax est vas in quo ex igni vivo calor ad materiam accommodatus, industriéque regitur" (ibid., p. 9).

112Ibid., pp. 22–23.

113"Elaboratio est operatio manuaria, qua res in substantia, eáque potissimum materiali mutata elaboratur" (ibid., p. 29).

114"Exaltatio est operatio qua res affectionibus mutata, ad altiorem substantiae & virtutis dignitatem perducitur" (ibid., p. 71).

by a variety of physical and chemical means. The whole proceeds through a maze of such dichotomized divisions, but once the process of definition and division is begun, it proceeds with its own inherent logic to exhaust the verbalized description of chemical operations.[115]

The dichotomies which Libavius employs in the disposition of the operations also underlie his distribution of chemical species in the second book of the *Alchemia*. This is not surprising, since the chemical species are regarded as products of these operations, not as unique material species to be classified in terms of their inherent chemical composition. The two parts of *alchemia* are linked as cause and effect. Chemistry for Libavius essentially involves the working up of some crude material substance in order to enhance its inherent properties and to free it from extraneous matter. What is most striking about Libavius's classification of chemical species, however, is his retention of the terminology of the alchemical, distilling, and Paracelsian traditions. Thus we find prominent in his scheme such entities as magisteries, quintessences, arcana, mysteria, and elixirs.[116] This only reflects, however, Libavius's respect for the traditional terms of the art. As he points out in several places in the *Alchemia*, these names have only a tropological or analogical significance.[117] The whole endeavor of the *Alchemia* is to define and distinguish definitively the species represented by these names from one another and thereby to identify them uniquely. In applying his dialectical dichotomies to the classification of chemical species, Libavius sought to bring discrimination to the symbolic and deliberately evocative names of the alchemical and Paracelsian vocabulary. The names remain the same, but their significations are sifted through the riddle of dichotomized definition, to emerge as discrete, differentiated clusters of homogeneous definables. Herein lies Libavius's definitive break with Paracelsian tradition. The open-ended book of signs is transformed into the discrete prosaic units of the

[115] For the relationship of these various categories see plate 2.

[116] It would require a detailed analysis of all the sources listed by Libavius to determine the influences on his classification of chemical species. But the affinity of his names of species with those described in Paracelsus's *Archidoxis* is unmistakeable. It will be recalled that Libavius viewed this work as Paracelsus's closest approach to a methodized chemistry. Cf. T. P. Sherlock, "The Chemical Work of Paracelsus," *Ambix* 3 (1948): 33–63; and R. Multhauf, "The Significance of Distillation in Renaissance Medical Chemistry," *Bull. Hist. Med.* 30 (1956): 329–46. The latter stresses, however, the unique significance of magisteries in Libavius's scheme.

[117] For instance, "Quae item species chymica his vocatur, tropo quodam etiam essentia nominatur, quanquam non desint qui arcana, & astra, & quovis grandi titulo quodcunque quovis modo praepaverint, appellare non erubescunt" (*Alchemia*, p. 86).

printed text. Where before there was echo, affinity, and sympathy now there is definition, division, and distinction.

The fundamental division of Libavius's chemical species is encapsulated in his definition of *alchemia* itself: it is the distinction between magisteries and extracted essences. This rests on the division between those species which derive from some extraction process (the category of *extracta*) and the magisteries, which are elaborated without any separation of essence from material matrix being involved.

The *extracta* are in the main preparations dependent upon such procedures as distillation, sublimation, crystallization, and precipitation. They involve the isolation of some inner essence of the starting material(s) separated from the grosser elementary parts.[118] Libavius's dichotomized arrangement of the various subspecies of *extracta* and essences proceeds in descending order of refinement and increasing order of substantiality. Thus the arcana are distinguished from the quintessences by the fact that they retain some specific character of the substance from which they are derived.[119] The astral arcana (tinctures and oils) retain the qualities of the elements air and fire, while the material arcana are related to the elements water and earth.[120] The "watery" arcana include distilled spirits and the various mineral acids;[121] the "earthy" arcana are comprised of alkalis and mineral species crystallized from solution, as well as solid sublimates and precipitates of an amorphous character.[122] Throughout the *Alchemia* can be discerned an attempt to classify the products of chemistry in terms of dichotomies derived from Aristotelian philosophy.

In contrast with the *extracta,* the magisteries are species which are perfected without the separation of any inner essence from the elementary matrix of the starting material. A magistery is defined as "a chemical species elaborated and exalted from the whole without extraction, the external impurities simply having been removed."[123] One characteristic feature of a magistery which Libavius emphasizes is that its preparation involves no significant loss of quantity or mass (*moles*)

118"Extractum est quod è corpore concretione, relicta crassitie elementari extrahitur" (ibid., p. 242).

119"Arcanum specificum est extractum naturae interioris, cuiusque speciei substantiam refereǹs proprius, ut in illa agnosci queat" (ibid., p. 336).

120Oils and tinctures are also classified as the "formal" arcana, as opposed to the "material" arcana (ibid., pp. 275 and 336).

121These are distinguished in terms of their dissolving and corrosive properties (ibid., pp. 336–69).

122Ibid., pp. 370–408.

123"Magisterium est species chymica ex toto citra extractionem, impuritatibus duntaxat externis ablatis, elaborata exaltataque" (ibid., p. 86).

apart from the small initial loss due to the cleansing process.[124] Magisteries are thus chemical species which retain their corporeal and elementary nature but whose properties or virtues are, as it were, concentrated and enhanced within their material substance. The Aristotelian distinction between quality and substance serves to dichotomize the category of magisteries. The magisteries of quality are exceedingly diverse. In the main they refer to various operations which enhance the physical properties of material substances, such as the finishing of metals and the cutting and polishing of gems.[125] They also include products of a change of state, such as the preparation of potable forms of the metals and the fixation of volatile species.

The magisteries of substance are divided into those of *genesis* and *catalysis*. The former can include a transformation of the whole substance, as in the transmutation of metals, or a combination of two separate substances to produce a new chemical species, as in the production of cinnabar from mercury and sulphur.[126] The magisteries of *catalysis*, on the other hand, are products of operations which resolve the whole into its parts.[127] Included in this category are the separation of metallic ores from their dross, the extraction of metals and minerals from their ores, and the separation of pure metals from alloys. The last magisteries described are illustrations of the separation of the four Aristotelian elements and of the three Paracelsian principles from a variety of starting materials.[128]

This inclusion of a discussion of both the four elements and the three principles at the end of the section on magisteries is indicative of what Libavius's chemistry is and, at the same time, what it is not. There is no attempt on Libavius's part to reconcile these two contending theories of matter: they are simply distinguished and bracketed as a dichotomy. They find their place in Libavius's scheme solely as part of the current rhetoric of chemistry, which must therefore be absorbed into the dialectic. Their unobtrusive presence in the middle of the work serves as a reminder that Libavius's *Alchemia* is not a treatise on matter and material transformation organized around theoretical principles. It is rather a systematized textbook of chemical operations and prescrip-

[124]"Unde & penè eadem relinquitur quantitas seu moles, quam natura per se dedit . . ." (ibid.). This characteristic is also specified in the definition of a magistery of substance (ibid., p. 130).

[125]Ibid., p. 90. Magisteries of quality also include magnetism (a magistery of occult quality), color, smell, taste, and sound, as well as physical changes of state.

[126]Ibid., pp. 130–204.

[127]Ibid., pp. 204–41.

[128]Ibid., pp. 228–41.

tions culled from the collective experience of all those who in their own way and for their own ends had practised and described techniques and preparations which Libavius regarded as chemical. Its model was dialectical and not axiomatic. Libavius, of course, like all Ramists and semi-Ramists, would not have acknowledged the difference. For them the process of methodizing a subject thereby made it a science and rendered unnecessary the search for first principles in an Aristotelian sense. Few perhaps would have been as fastidious as Libavius was in his chemical epistles in attempting to reconcile the methodized art with the canons of Aristotle's *Analytics*. But in this strange pedagogical world, art, science, and method all merged and found their common denominator in anything that could be taught. Libavius called chemistry into being simply by demonstrating that it was teachable.

The ideals which informed Libavius's contribution to chemistry were thus deeply rooted in the humanist pedagogical theory of the sixteenth century. He took the program initially formulated for training in eloquence and applied it to the realm of chemical technology. Out of the techniques and prescriptions of the previously distinct chemical crafts and traditions he methodized a dialectic of chemistry. His object in so doing was to make chemistry nothing more nor less than a curriculum subject. By making it an art which could be taught, he sought to bring it into the public domain in a form suitable for institutionalization. It was only then, he believed, that its fruits could truly benefit society. Libavius's *Alchemia* remains as a remarkable and extraordinary monument to sixteenth-century Lutheran humanism.

CHAPTER VII

EPILOGUE

This book has been a study in the contrast offered by two prominent chemical writers at the turn of the seventeenth century. On the one hand was Oswald Croll, as representative and spokesman for the enthusiastic alchemical ideology of the Paracelsians; and on the other, Andreas Libavius, a feisty defender of orthodox religious humanism who sought to make chemistry a methodized didactic. The all-embracing chemical world view of Croll, which tails off in his text into silence, is pitted against the all-embracing textbook of Libavius, which exhausts the techniques and prescriptions of the chemical arts in a welter of wordy definitions. The object of this study was to illuminate the relationship between the Paracelsian chemical philosophy and the later textbook tradition of chemistry, which was such a marked feature of its seventeenth-century development. It has been found that this textbook chemistry does not simply evolve from Paracelsian philosophy by the abstraction of the positivistic elements of Paracelsian prescription and the abandonment of its more mystical elements. Rather it was the product of a fundamental clash of cultural ideologies which had much deeper significance and import than the nature and provenance of chemistry. Ultimately this conflict centered on the origin, nature, and signification of words and language. Was the word a vibrant echo of the divine Word of creation, which reverberated through the whole of nature in manifest forms not only verbal? Or did there exist only the imperfect utterances of men, which called experience into being and which through an historical process of refinement and conscious criticism organized the totality of collective experience? Oswald Croll invites us to step out of his book and experience immediately the Book of Nature; Andreas Libavius attempts to condense chemical knowledge and present it to us as a logically ordered arrangement of printed words. For the former, chemistry was to be experienced; for the latter, methodized.

These two outlooks were rooted in different aspects of the Renaissance revival of ancient knowledge. The Crollian view was linked to the Hermetic vision of ancient wisdom, which can be traced back to the literary endeavors of Ficino and Pico della Mirandola. This view saw the writings of antiquity as veiled repositories of some *prisca sapientia* which it sought to recover and put into practice in mystical and magical rites. The tradition to which Libavius belonged was of a more orthodox humanist variety which saw in the writings of antiquity the most eloquent memorials of right reasoning and conduct, whose texts only required interpretation and clarification. The one sought to divine the meaning behind the textual word; the other sought simply to collate, to compare, and to criticize.

Chemistry entered the picture through the association of Paracelsian chemical naturalism with the Hermetic tradition. In this form the literary origins of Hermeticism receded, and the Book of Nature took primacy over the books of men. Out of this alliance grew a radical gnostic philosophy and theology which challenged orthodox learning and all its institutional forms. It was the enthusiastic and revolutionary character of Paracelsian ideology, with its essentially chemical view of the world, which Libavius sought to counter by demonstrating that there was nothing transcendent or mystical in chemistry and that chemical technique and practice were amenable to methodization by application of the rules of dialectic. By articulating chemistry in the textbook, Libavius's aim was to give it a format which could be utilized within the existing institutions of learning. He thereby sought to undercut its revolutionary implications and associations.

Libavius's invention of chemistry as a discipline was to have critical influence on the development of the subject in the seventeenth and eighteenth centuries. The fact that chemistry emerges first as a didactic may help to explain some of the curious features about its history in this period. If there is one tangible form of chemistry at this time it is the didactic form. The seventeenth-century chemical literature abounds in teaching manuals, many of which were related to actual courses of instruction both private and institutionalized. The most popular of these, like Jean Beguin's *Tyrocinium Chymicum* (first edition Paris, 1610) and Nicolas Lemery's *Cours de Chymie* (Paris, 1675) enjoyed spectacular success and ran to numerous editions. There was almost a continuous tradition of chemistry courses in Parisian scientific life throughout the seventeenth and eighteenth centuries, and they became institutionalized at the Jardin Royal des Plantes from 1648 on.[1] The teachers of the majority of these courses would produce their

[1]The most thorough discussion of the French textbook tradition is Hélène Metzger, *Les Doctrines Chimiques en France* (1923; reprint ed.; Paris: Albert Blanchard, 1969). The history of chemical instruction at the Jardin Royal des

own textbooks. In Germany and Holland, chemistry even found its way into the university curricula. Courses and textbooks appeared from Jena, Altdorf, Utrecht, and Leyden.[2] This is a remarkably rapid assimilation of a new practically oriented discipline into what is usually regarded as the refractory environment of the seventeenth-century university. And all of this in relation to a discipline which can hardly be regarded as having attained intellectual maturity in this period. It appears that we cannot underestimate the self-generating power of didactic.

While the detailed influence of Libavius and his *Alchemia* on the subsequent textbook tradition has yet to be fully explored, certain relationships are already clear. It has been shown that Jean Beguin, the fountainhead of the French chemical didactic, lifted verbatim key definitions from Libavius's *Alchemia* in the first edition of his *Tyrocinium Chymicum.*[3] This work for the chemical beginner would appear to be Libavius's idealized didactic reduced to manageable form for use in an actual course. The more than fifty editions and translations of this elementary text would sustain the Libavian ideal well into the second half of the seventeenth century. In 1630 Zacharias Brendel at Jena would give a belated response to Libavius's epistolary invitation to his father to methodize the chemical art by publishing a textbook for his course significantly entitled *Chimia in Artis Formam Redacta.*[4] This text would again appear to be in part Libavius simplified, being organized around the four degrees of heat which Libavius described in the *Pyronomia* section of the *Alchemia.* This textbook was reissued in several editions subsequently, to be superseded by Werner Rolfinck's text of the same title in 1661.[5] In general the texts which derive from university instruction in Germany and Holland in the course of the seventeenth century are fuller and more pedantic (academic) in their

Plantes is dealt with in J. F. Contant, *L'Enseignement de la Chimie au Jardin Royal des Plantes de Paris* (Cahors, 1952).

[2]No systematic study of chemical instruction in the seventeenth-century universities has been published, but a preliminary survey was made in the author's *Early University Courses in Chemistry* (Ph.D. diss., The University of Glasgow, 1965).

[3]A. Kent and O. Hannaway, "Some New Considerations on Beguin and Libavius," *Ann. Sci.* 16 (1960): 241–50.

[4]Zacharias Brendel, *Chimia in Artis formam redacta* (Jena, 1630). Subsequent editions appeared in 1641 Jena; 1659 and 1668 (Amsterdam); and 1671 (Leyden).

[5]Werner Rolfinck, *Chimia in Artis formam redacta, sex libris comprehensa* (Jena, 1661). Of all the seventeenth-century textbooks, this one most closely resembles that of Libavius in form; it even retains the Ramist dichotomized tables.

treatment of chemistry than the series of French manuals. But all have a common form of organization: the definition of the art, a description of its instruments, a discussion of operations, followed by preparations—that is, the basic structure of the *Alchemia*. It is this fundamental dialectic of technique and prescription which gives identity and continuity to chemistry throughout the seventeenth and into the eighteenth century. While the authors of these textbooks might incorporate different theoretical principles and explanatory systems to interpret chemical reactions, rarely do these ever alter this basic structure of presentation. Furthermore, as many of these texts make explicit in their varying definitions: "Chemistry is an art which teaches. . . ." Some would be more concise and simply call chemistry an art; others would call it a scientific art, or an art and a science, thus laying a trap for the unwary who would later debate as to whether chemistry in the seventeenth century was indeed a science or simply applied empirical knowledge.[6] But in the dialectical and pedagogical theory in which chemistry had its roots, art and science simply signified anything that was teachable.

It is as an established didactic that chemistry passes into the Age of Enlightenment: a methodized discourse, conscious of its curricular identity but still in search of its own theoretical principles and problematic. Its most important and influential academic spokesman of the early eighteenth century, Herman Boerhaave, would offer the following definition of the subject, which has more than one echo of Libavius:

> Chemistry is an art which teaches the manner of performing certain physical operations, whereby bodies cognizable to the senses, or capable of being rendered cognizable, and of being contained in vessels, are so changed by means of proper instruments, as to produce certain determined effects, and at the same time discover the causes thereof; for the service of various arts.[7]

The subsequent history of chemistry in the eighteenth century is largely the search of this art for an organizing explanatory theory. The phlogiston theory of combustion and the affinity theory of chemical reaction would both partially supply that lack, but never wholly successfully. Chemistry would not attain scientific maturity until Lavoisier

[6] For example: "Alchymia est ars quae purum ab impuro separare docet" (Jean Beguin, *Tyrocinium Chymicum* [Paris, 1610], p. 1); " . . . la chymie est un art scientifique qui enseigne . . . " (Christopher Glaser, *Traité de la Chymie* [Paris, 1663], p. 6); and "La Chymie est un Art qui enseigne . . . " (Nicolas Lemery, *Cours de Chymie*, 2d ed. [Paris, 1677], p. 2).

[7] Herman Boerhaave, *A New Method of Chemistry . . .*, trans. by Peter Shaw, M.D., 2 vols., 2d ed. (London, 1741), 1: 65; reproduced in J. R. Partington, *A History of Chemistry*, 4 vols. (London: Macmillan, 1961–70), 2: 746.

reorganized it around his oxygen theory of combustion. Many factors outside the strictly didactic tradition would contribute to Lavoisier's chemical revolution, but that revolution was not completed until it was codified in a textbook. Against the background of the sixteenth-century textbook origins of chemistry, Lavoisier's *Traité Elémentaire de Chimie* acquires a fresh significance. In this text, designed specifically to win over the minds of youth to the new chemistry, Lavoisier employs the pedagogical and linguistic theories of the Abbé de Condillac to systematize anew the science. It is through a new theory of language—one that does not simply permit words to discriminate chemical species from one another but which gives words the power to penetrate the substance of chemical entities and to analyze them—that Lavoisier would write another chapter in the story of the chemists and the word.

INDEX

THE JOHNS HOPKINS UNIVERSITY PRESS

This book was composed in Baskerville text and Albertus display type by The Composing Room of Michigan, from a design by Patrick Turner. It was printed on 60 # Clear Spring Book Offset paper by Collins Lithographing and Printing Co. and bound in Holliston Roxite cloth by Haddon Bindery, Inc.

Library of Congress Cataloging in Publication Data

Hannaway, Owen.
The chemists and the word.

Includes bibliographical references and index.
1. Chemistry—History—Sources. I. Title.
QD11.H27 540'.9 74-24380
ISBN O-8018-1666-1